The Insect-Populated Mind

How Insects Have Influenced the Evolution of Consciousness

David Spooner

Hamilton Books
A member of
The Rowman & Littlefield Publishing Group
Lanham • Boulder • New York • Toronto • Oxford

Copyright © 2005 by
Hamilton Books
4501 Forbes Boulevard
Suite 200
Lanham, Maryland 20706
Hamilton Books Acquisitions Department (301) 459-3366

PO Box 317
Oxford
OX2 9RU, UK

Library of Congress Control Number: 2005925771
ISBN 0-7618-3175-4 (paperback : alk. ppr.)

∞™ The paper used in this publication meets the minimum
requirements of American National Standard for Information
Sciences—Permanence of Paper for Printed Library Materials,
ANSI Z39.48—1992

In Memory of Norman O. Brown
whose associative method and encouragement
inspired this book

Contents

Contents

Foreword

For the independent scholar, nothing is as important as Libraries and their librarians. I have been fortunate enough to benefit from a number of excellent Scottish libraries. St. Andrews University has an unmatched collection of classical texts and journals; Stirling University provided many of the basic texts I analyze; Edinburgh University holds a Special Collection of Charles Bonnet works along with the Alfred Russel Wallace Collection; and the staff at the National Library of Scotland have been ever resourceful. I must also thank the Library of Congress and the British Library for the transmission of a number of key documents. The Henry W. and Albert A. Berg Collection at the New York Public Library kindly supplied two unpublished Thoreau documents on Insects. They were skillfully deciphered by Barbara Berliner, to whom sincere thanks.

Some of the chapters have previously appeared elsewhere. I am grateful to the publishers and editors for making them available again.

"From Apuleius to A.R. Wallace: Evolutionary Theory and some literary animals and insects" was delivered at the Venezuela Congress of the Beast Fable Society, and was published in *Bestia* 4 (1992).

"Of Cells and Mutation" was given at the Glasgow Beast Fable Congress under the auspices of Glasgow University and Northeast Missouri State University, and appeared in *Bestia* 5 (1993).

An early version of "Shakespeare and the nature of verbal and musical language" was published in the author's *The Metaphysics of Insect Life and other essays* (San Francisco, 1995).

"Thoreau's Insect Science" appeared in *Thoreau's Vision of Insects & the origins of American Entomology* (Philadelphia, 2002).

"Insects and Mind: the poetry of Damaso Alonso" was first published in *The Poem and the Insect: aspects of twentieth century Hispanic culture* (Lanham, Md., 1999 & 2001).

The author is pleased to acknowledge the inclusion of brief quotations from Nabokov's works, the sources of which are detailed in the bibliography. All quotations are from Penguin editions, except where stated.

I have been encouraged over a long period by a number of longstanding colleagues and friends: the late Peter Mann, a scholar of the most uncompromising integrity from the School of English, Leeds University; Dr. Keith Carabine, Chair of the Conrad Society and scholar of immense intellectual generosity; Jeffrey Wainwright, poet and critic; and the late David Goodman, President of the International Brigade Association and Warden of Wedgewood College, who was a great representative of the universalist outlook and intellect of the volunteers in the Spanish Civil War.

Throughout my writing career my wife, Marion O'Neil, has been prepared to interrupt her work as Principal Illustrator at the National Museum of Scotland and sustain the positive spirit of the venture.

David Spooner
Dunfermline, Scotland
January 2005

"We seek to decode nature's gigantic cryptogram in such a way that structures emerge which are conserved under various changes and metamorphoses."

—Leibniz

"The important things in the world appear as the invariants of transformations."

—Dirac

"[Falter] pointed out that, having accidentally solved 'the riddle of the universe,' he had yielded to artful exhortation and shared that solution with his inquisitive interlocutor, whereupon the latter had died of astonishment."

—Nabokov

"There lie hidden in language elements that effectively combined
Can utterly change the nature of man."

—MacDiarmid

"And Coleridge, too, has lately taken wing, But like a hawk encumbered with his hood,—Explaining Metaphysics to the nation—*I wish he would explain his Explanation!*

—Byron

Part One

Chapter One

Ape, Insect and Human

The relation between humans and the great apes has been exhaustively investigated over the past 40 years. But it has been entirely overlooked that there is a very close connection between aspects of insect evolution and the human intellect. Alfred Russel Wallace, the co-founder with Charles Darwin of the theory of evolution by natural selection, threw down the intellectual challenge over a hundred years ago. He wrote that while the human body "was undoubtedly developed by the continuous modification of some ancestral animal form, some different agency, analogous to that which first produced organic *life*, and then originated *consciousness*, came into play in order to develop the higher intellectual and spiritual nature of man."[1] My book will move between entomology, language theory, genetics, astronomy, literature and music. It will show how these apparently disparate subjects are fused, and that this fusion tells us, at the last, some crucial things about human nature in its determining association with insects. There has been little progress in such cross-disciplinary arenas for reasons spelt out by E.O. Wilson in relation to the field of biology:

"If human behavior can be reduced and determined to any considerable degree by the laws of biology, then mankind might appear to be less than unique and to that extent de-humanized. Few social scientists and scholars in the humanities are prepared to enter such a conspiracy, let alone surrender any of their territory."[2]

There is a wary truce between science and art for the moment, but with continual advances in astronomy and in the subjects that most nearly affect the human body and health such as genetics, this is not destined to last for long. Drawing together science and the humanities (or more abstractly object and subject) is not a question of $a + b$, but of $a \times b$. Then what one arrives at

3

is not recognizable as either science or art, *a* or *b*, but represents *ab*, and offers insights into human nature that the most determined fact-finder cannot afford to overlook.

In one of the papers that inaugurated cognitive psychology, George A. Miller wrote of the numeral 7:

> "My problem is that I have been persecuted by an integer. For seven years this number has followed me around, has intruded in my most private data, and has assaulted me from the pages of our public journals. This number assumes a variety of disguises, being sometimes a little larger and sometimes a little smaller than usual, but never changing so much as to be unrecognizable. The persistence with which this number plagues me is far more than a random accident."[3]

Substitute the number four for Miller's seven, and slightly adjust the wrapping, and my Kafkaesque situation is perfectly described. Vladimir Nabokov in his story "A Busy Man" describes how the number thirty-three "had got entangled with his unconscious, its curved claws like those of a bat, had got caught in his soul, and there was no way to unravel that subliminal snarl."[4] So it is when a certain number seems to haunt many aspects of life, and to interlink the observer and the thing-in-itself. Schopenhauer concludes the first book of Volume 1 of *The World as Will and Representation* with the observation that a division into the perceiving subject and an external object leaves us eternal watchers. Nothing happens, and "we can never get at the inner nature of things *from without*."[5] Schopenhauer then draws the critical conclusion that the inner nature of the object is at the same time the inner nature of the self. A truly dynamic theory of knowledge should reveal this inner connection which raises both object and subject from the status of mere phenomena.

How does the artisan strength of Blake's Fourfold Vision relate to the intricacies of Relativity and Quantum theory on the one hand, and Shakespeare's and Melville's works on the other? While not disputing the existence of complexities of organization in organisms, why should we so worship complexity and indeed size when the DNA coding in some species of amoeba is 30 times greater than in humans? Leaving aside the remarkable closeness of the bonobo or pygmy chimpanzee's genome to our own, even daffodils have 35% of the human genome! We have, I believe, become over-tortuous in our celebration of complexity for its own sake. "Simplify, simplify," wrote Thoreau—sensing the need for a seriously emphatic "Simplify," while Augustin Fresnel explained that "simplicity lies concealed in this chaos, and it is only for us to discover it!"[6] Occam's Razor, the so-called 'principle of parsimony,' lay behind Einstein's attributed epigram that "everything should be made as simple as possible, but not simpler."

And nothing is more simple than the instinctual numeracy of *homo sapiens*. When Martin Rees asks: "is there any particular reason why a universe should end up with *four* 'expanded' dimensions (time, plus three dimensions of space), rather than some different number?"[7] then the answer may be more straightforward than appears at first sight. It may be related to the fact that numerosity—the ability to estimate numbers of items without actually counting—reaches its limits at four, and that this is an aspect of humanity's at-homeness in the cosmos by whatever chance or incidental physical mode this may have occurred. What I intend to ask throughout this book is a series of queries that have been elbowed to one side in recent times: how 'at home' in the universe should the human race reasonably feel? As Emerson put it: "External nature is only half. The geology, the astronomy, the anatomy are all good, but 'tis all a half, and—enlarge it by astronomy never so far—remains a half."[8] What I try to show is that, working within an aegis of the Weak Anthropic Principle, our cosmic environment is in tune with key structures in the internal life of humanity, that is the arts and religion. Otherwise, after all, the universe is reduced to "a dead lump here, a ball of gas there, a bit of fume somewhere else. How puerile it is, as if the universe were the back yard of some human chemical works!"[9] The vast diversity of the cosmos and of life on earth contains a skeletal backbone of structure that once seen is unmistakable, recurring in arena after arena. This goes beyond merely defining the physical, chemical conditions necessary to create 'observers' in the first place. While of course there is a massive amount of random numerical and physical activity in the universe, certain significant and recurrent patterns are present which fuse us with our cosmic surroundings. In the beginning, Indo-European language shows the centrality of the number four in origins.[10] Its roots suggest that it was a base-4 language in which the fundamental number was 4, arising most probably from our mineral- and chemical-built digits. So counting worked on the principle of $4 + 1, 4 + 2$ and so forth. The word for 8 in proto-Indo-European consists of a double form of the word for 4, that is $4 + 4$. Then in leaping to the word 'nine' a root seems to be 'new' as in German *neu* and *neun*, Italian *nove* (new) and *nuove* (nine), French *neuf* (nine and new).

Einstein elucidated a method applicable to all fields of intellectual endeavour:

"The scientist has to worm these general principles out of nature by perceiving certain general features, which permit of precise formulation, in large complexes of empirical facts. Once this formulation is successfully accomplished, inference follows inference, often revealing relations which extend far beyond the province of reality from which the principles were originally drawn."[11]

This will by no means be a book merely about numbers though, for it would be retrograde to end up back in the Victorian and post-Victorian competition between quaternions and octonions. Much earlier words and numbers were interconnected. The Judaic *gematria* drew on the fact that each Hebrew word was assigned a numerical value. If the sum of the letters of one word equalled that of another, then those words were reckoned to be associated in some way. This has been dated to 200 CE, but it has its origins in Babylon where a gematria representing the value of Sargon of Akkad's name was projected in the building of a wall in Khorsabad. However our concern here is rather with tangential connections between numbers, words and nature. Key clusters of words reach deep back into the formation of language, and represent both a response to surrounding nature, and some inner pivot that connects us not only to other primates, but to the insect. It is interesting initially to observe some general correlations between insect and number. The Gaelic word for butterfly is *dealbhan-dé*, meaning 'fire of God,' while the Danish word for 4 also signifies 'fire'. The Roman numeral 4—symbol of the divine Fire—is the Indian letter, *ch*, the initial of *chatur*, the Sanskrit for *four*. Persun 4 is, similarly, *chechar*.[12] A Nahuan butterfly is 'the hand of fire,' and indeed stands for fire via the flickering of flames, symbolized as an incense burner. The fire equivalence I interpret as indication of the urgent centrality of these concepts. The Persian *chahar bagh* signifies the fourfold garden, which was adopted by the Muslim invaders and incorporated into the Koran. Four rills flow into a central fountain and divide the garden into four quarters. On a different tack and another insect—the Egyptians elevated the scarab beetle to a symbol of Being itself, and the word for the scarab, *Chepera*, was also the word for Being.

Could the recurrent fourfold shapes and units be termed memes though? I am skeptical as to the overall Dawkins theory of memetics, and in this regard I shall be defining what I term the Cosmic Cultural Faculty (CCF), which I hypothesize as analogous to the multi-faceted organization of neural circuits that constitute the faculty of physical vision. In this analysis, the CCF rather appertains to internal vision, or spiritual perspective, and has its biological roots in logico-mathematical symmetries that run through the nervous system. It has been suggested that musical composition, with its evocation of the passing of time which is yet brought to closure by the formal structure of music, may offer some clues here. I will be suggesting in Chapter 3 that the sense of time elapsing as experienced in listening to (the usual) 4 movements of a symphony or quartet is also a passing through 4 major stages of intellectual and spiritual life corresponding to the 4 stages undergone by those insects that undergo complete metamorphosis. In other words, the key works of

Beethoven, Mozart, Haydn, and Brahms, suggest an especially exact resemblance between music and our sense of time passing, of *growth*. Indeed they are a clue to the inner meaning behind language—quite beyond the structural grammars essayed in Terrence Deacon's epic *The Symbolic Species*, or in Chomsky, central though those are to everyday communication.

An all-round conspectus of the greatest of human achievements through the centuries in the arts and religions makes it clear that Alfred Russel Wallace's interpretation of natural selection is nearer to the truth of the human condition than Darwin's. He has of course been widely criticized for being too inflexible in his understanding of evolutionary theory. It is this simplistic method—so it is argued by Stephen Jay Gould, John Maynard Smith and John D. Barrow—that prevented him from seeing that the mind could be a product of evolution by natural selection.[13] The failure was, according to Barrow, an inability to see that "we are bundles of abilities, outdated adaptations, and unconscious by-products."[14] However Wallace had good reason not to be diverted from his scepticism by these contingent qualifications, and anyway these aspects are only circumstantial. There is a double evolutionary relationship at work in the human experience. On the one hand, there is a direct physiological one with the great apes, but on the other there is a latent and implicit one with insects through transformations of the intellect. Nuances become, at a certain point, crucial. None of Wallace's writing matched the first great statement of Darwin's *The Origin of Species* in showing how blind processes of random variation and natural selection combine to provide a counterfeit and misleading 'design' to evolution. But other, apparently subsidiary themes of Wallace such as his great theme of mimicry in butterflies also have a indicative role to play. The contemporary of Wallace and specialist entomologist, Henry Walter Bates, was the first to articulate their key role in evolutionary theory:

> "It may be said, therefore, that on these expanded membranes Nature writes, as on a tablet, the story of the modifications of species, so truly do all the changes of the organisation register themselves thereon. And as the laws of Nature must be the same for all beings, the conclusions furnished by this group of insects must be applicable to the whole organic world; therefore the study of butterflies —creatures selected as the types of airiness and frivolity—instead of being despised, will some day be valued as one of the most important branches of biological science."[15]

Here the writer Nabokov, Bates and Wallace have much in common, as I show in Chapters 4, 5 and 14. Something of the excitement of approaching evolutionary theory by way of the insects is caught in Nabokov's poem about his discovery of the Karner Blue butterfly, which parallels the rapture expressed by Wallace in his capture of the *Ornithoptera croesus*:

> "I found it and I named it, being versed in
> taxonomic Latin; thus became
> godfather to an insect and its first
> describer—and I want no other fame."[16]

While the Cosmic Constant Faculty (CCF) contributes to the fitness of the individual, it is doubtful whether it would have any perceptible impact on genetic organization. However, it is not entirely impossible that over a lengthy period, the CCF has some epigenetic significance. In that respect, the faculty of fourfold cultural structure has to await any direct evidence from the brain sciences and genetics one way or the other. Deciding where on a line between programmed and learned an aspect of the faculty lies is often not possible. So Brent Berlin and Paul Kay have waited over thirty years for proof one way or another that the similarity of words for colour in a multitude of languages is a universal trait of neurophysiology. An informed guess though would be that the CCF is part of biological endowment, rather than any physically quantifiable characteristic. However discoveries involving the glia cells in the brain over the past few years suggest that these cells which were thought to be merely a maintenance feature, but which process memory, could open the way to understanding more thoroughly the way that the mind absorbs and processes cultural stimuli. It transpires that neurons and glia carry on a two-way dialogue from embryo to old age. A contemporary American writer, Bob Hicok, who has been described as "the nation's most prolific nonacademic serious poet since William Carlos Williams," has said that "I have to believe that writing is simply in me. . . . I've come to believe the desire [to write] is biological, that there is no reason other than this in what I was going to do."[17] As Luigi Cavalli-Sforza and Marcus Feldman conclude in relation to cultural traits, "it is difficult to partition the process of transmission into purely genetic and purely cultural components," but they are passed on by "imprinting, conditioning, observation, imitation, or as a result of direct teaching", and not merely by imitation as demanded by the theorists of memes.[18] The primary 'motives' that determine this particular biological process are the quest for equilibrium, the product of continuous tension between conservation and entropy. This is akin to the universal state elaborated by Barrow and Frank Tipler:

> "The sizes of stars and planets, and even people, are neither random nor the result of any Darwinian selection process from a myriad of possibilities. These, and other gross features of the Universe are the consequences of necessity; they

are manifestations of the possible equilibrium states between competing forces of attraction and repulsion. The intrinsic strengths of these controlling forces of Nature are determined by a mysterious collection of pure numbers that we call the *constants of Nature*."[19]

I am working from the inside (of literature, music, religions) outwards, so I am seeking structures that are common to all three, but which are yet found prolifically in the cosmos. Disapproved of by most physicists, the word 'mind' has come into play in this book, specifically as a product of the instinct toward Cosmic Cultural Constants. This is not, of course, to deny or belittle the constants of nature. But it is to suggest that being-at-home in the universe for *homo sapiens sapiens*—and let us not forget our erstwhile cousin *homo sapiens neanderthal* as well—dialectically transcends the knowledge of those constants. Which is to say it includes but ranges beyond that knowledge.

The CCF appears to have something of the character of the Lumsden-Wilson "culturgen"—that is it seems "the basic unit of inheritance in cultural evolution."[20] Even Noam Chomsky in regard to his pioneering concepts sees as "a distant prospect for inquiry," any genetic connection to Universal Grammar and the Language Acquisition Device.[21] A similar situation holds for the search for a genetic basis for the Number Module, our instinctive ability to handle numbers.[22] The Cosmic Cultural Faculty is rather an assemblage of biological processors that has enabled inspired intuitive individuals to create art, and communities to project religions characteristically built upon structures that frequent the cosmos. The faculty is probably adaptive in the Darwinian sense in increasing an individual's chances of survival success, perhaps most poignantly exemplified by the endurance of some musicians in the Nazi concentration camps.

My theory of patterns of intellectual and spiritual maturation reaching back to the processes of metamorphic insects works in tandem with the demands of Darwinian evolution, being dependent upon variation and the ability to replicate, together with interaction of the specific originality with the historical environment. The contention is that the indisputable primate connection has caused mainstream evolutionary theory to miss the all-round interrelationship of human development to entomology, and that this relation is enshrined in the greatest of the higher art forms and religion. There is a crucial oblique relationship between metamorphic insects and humans, a connection transmitted through the great works of music and literature, and through many of the paradigms of world religions. At the outset I should say that the concentration is on the metamorphoses of Lepidoptera, the butterflies and moths. The four orders of higher insects are the only ones with perfect metamorphosis—Lepidoptera, Coleoptera (beetles), Hymenoptera (ants, bees and wasps) and Diptera (two winged flies, mosquitoes etc.). The Hymenoptera constitute a primarily eusocial class. E.O. Wilson has defined the source of their sociality:

"The tendency of aculeate Hymenoptera to evolve eusocial species can probably be ascribed in part to their mandibulate (chewing) mouthparts, which lend themselves so well to the manipulation of objects, or to the penchant of aculeate females for building nests to which they return repeatedly, or to the frequent close relation between mothers and young. These and perhaps some other biological features are prerequisites for the evolution of eusociality."[23]

Lepidoptera evince quite different behaviour which could be characterized as more individually orientated—with the obvious proviso that the foodplant needed by the larvae has to be taken into account.

Indeed the underestimation of the link between intellectual and spiritual processes with the world of insects may well be unwittingly encouraging the sixth mass extinction now taking place. This finds the insects comprising half the world species under particular pressure, and in comparison with the mammals, relatively unappreciated except by specialists. 90% of insects have yet to be described. Insects are unmatched barometers of the health of an ecology, almost immediately registering adverse changes in the environment through their symbiotic relation with earth. In relying upon them, we are calling up the evidence of one of the earliest stages of life on earth for our information and forewarnings of change. E. O.Wilson in his tireless fusion of biology and ecological defence has ingeniously coined the term 'biophilia' to indicate our need to affiliate to other forms of life. In Britain, as the latest surveys reveal, "a majority of [British] butterfly species (71% over ~ 20 years) has declined," and the extinction has been two orders of magnitude higher than for plants and vertebrates.[24] As the great zoologist Friedrich Christian Lesser put it, "Les Insectes se conduisent selon les règles de la sagesse."[25]

Although a theory of everything is usually regarded as the specialist scientist's Holy Grail—or maybe mere will o' the wisp—being the unification of the macrocosmic Relativity and the microcosmic Quantum mechanics, it is far more compellingly thought of as a theory that not only explains how our universe came into being, but why for the moment at least it is the only one there could have been, and how it came to include one planet where human life could exist. (This is not to rule out another galaxy with a different sun, and another planet with some form of life, perhaps purely electronic.) Such a general theory cannot avoid seeking a relation between spiritual symmetries on the one side, and the nature of physical laws on the other. Even if one, in post-Heideggerian fashion, name-calls it a metaphysic nonetheless in that it will unify science and the super-structural arts, it will reveal entirely new perspectives on the place of the human species in nature. And this is no straight-forward matter, since it affects how we view the nature of the human species and its place in the universe.

As with the place of the individual in Relativity and Quantum mechanics, a theory of everything must place subjective human achievement in sharper focus than at any time since Galileo, and inevitably draws in Beethoven and Dante as much as Einstein or Planck. Steven Weinberg in his *Dreams of a Final Theory* argues that a comprehensive knowledge would have a far greater impact on human psychology than on human technologies.[26] This should come as no surprise since modern science has put the observer and their perspectives under the closest scrutiny. As Paul Davies has remarked, "the nature of reality and perhaps the very structure of the universe is intimately related to our existence as conscious individuals perceiving the world around us."[27] Although Artificial Intelligence promised much, it has collapsed into a form of technical solipsism. The essential route towards an all-encompassing abstraction brings together the more nearly objective sphere (the physical sciences) with the more subjective spheres (arts and religion), via the mediation of evolutionary theory. Such a synthesis will prove uncomfortable, overturning previous assumptions and seriously challenging the divisions of intellect enshrined in the academies. Schopenhauer wrote, perhaps a little unfairly to the poets:

"Thus whereas the works of poets pasture peacefully side by side like lambs, those of philosophy are born beasts of prey and, even in their destructive impulse, they are like scorpions, spiders, and the larvae of some insects and are turned primarily against their own species. . . . The philosopher's work . . . tries to revolutionize the reader's whole mode of thought. It demands of him that he shall acknowledge as error all that he has hitherto learnt and believed in this branch of knowledge; that he shall declare all his time and trouble to be wasted; and that he shall begin again at the beginning. At most, it leaves standing a few fragments of a predecessor in order thereon to make its foundation."[28]

The sheer interrelatedness of the crisis in all areas of human thought means that to re-interpret one field is to involve all other fields in that crisis. I shall be applying Occam's razor to a vast swathe of data and commentaries in order to open up new perspectives on the problems. The initial approach is through the sphere of music, literature and language, for as the word imagination suggests by way of its associative root 'imago', the natural world is inextricably associated with the workings of the creative aspect of the mind. The Fibonacci sequence, which reveals the remarkable symmetry of flowers by a numerical succession that codes the petals progressing through sums of previous numbers—1, 2, 3, 5, 8, 13, 21, 34, 55, 89—is an index of the persistent and mysterious order of nature, as of the human mind. Likewise the cellular cleavage of one of the few non-insect metamorphic creatures—the frog—has its doubling sequence: 2, 4, 8, 16, 32. Elsewhere, the Golden Ratio is no longer the prerogative of the world of art, but is to be found in the arrangement of leaves on plant stems, the shape of a sunflower's seedhead and a seashell's spiral, and even to be found in the properties of spinning black holes. Dirac, enumerating constants in nature, specifies K_c/e^2 as one fundamental to atomic theory, and as having the value of about 137. This has a close relation to the value of the Golden Ratio.[29]

Perhaps the unfashionable poet Robert Browning sensed some finer cosmic symmetry, albeit cast by the random workings of natural selection, when he created the ambiguous Fifine whose medium is flowers, and who personifies the integrated life the poet seeks but is denied by his own complacencies, scepticism and conventionality. The anthropic principle in all its ramifications, the manifold coincidences of structure that make life on earth possible, also makes it in many ways beautiful. Or in Brandon Carter's famous definition—"our location in the Universe is necessarily privileged to the extent of being compatible with our existence as observers."

NOTES

1. Wallace (2), 2:17. Wallace's emphases.
2. Wilson, E.O. (1), 13,
3. Miller, 81.
4. Nabokov (14), 287.
5. Schopenhauer (5), 1:99. Schopenhauer's emphasis.
6. qd. Moszkowski, 45.
7. Rees, 144.
8. Emerson, 165.
9. Lawrence (1), 53.
10. The remainder of this paragraph is drawn from Butterworth, 66–67.
11. Einstein (2), 128.
12. Bayley, 2:4.
13. Smith (2), 94–95; Gould (3), 44–51; Barrow (1), 30.
14. Barrow (1), 30.
15. Wallace (7), 277–78.
16. Nabokov (11), 155.
17. Deutsch, 41, 42.
18. Cavalli-Sforza, 8, 7.
19. Barrow and Tipler, 5. Barrow and Tipler's emphasis.
20. Lumsden, x.
21. Chomsky, 4.

22. Butterworth, 161.
23. E.O. Wilson (2), 415.
24. Thomas, 1880.
25. Lesser, 298.
26. Weinberg, 192.
27. Davies (1), 183.
28. Schopenhauer (3), 2:5–6.
29. Dirac (1), 73.

Chapter Two

The Cosmic Cultural Faculty
in Relation to the Meme Machine

The category of memes has been developed by evolutionary psychologists as an explanation of Darwinian evolution in the sphere of culture. They are units of cultural expression that are capable of being imitated, along the lines of DNA, and so capable of replicating with very occasional errors of transcription. It is argued that just as genes are invisible, so memes are invisible and transmitted through "pictures, books, sayings."[1] Genes are encoded in molecules of DNA; memes arguably likewise in the human brain and thence into books. As Darwinian elements, memetics sees ideas as self-replicating independent forces, concerned only to be successfully copied:

> "DNA is a self-replicating piece of hardware. Each piece has a particular structure, which is different from rival pieces of DNA. If memes in brains are analogous to genes they must be self-replicating brain structures, actual patterns of neuronal wiring-up that reconstitute themselves in one brain after another."[2]

This is from *The Selfish Gene* (1976), where Richard Dawkins describes memes as "tunes, ideas, catch-phrases, clothes fashions, ways of making pots or building arches." These leap, as do ideas, from brain to brain in a process "which, in the broad sense, can be called imitiation."[3] However it is surely hardly necessary to create a whole brain structure merely for a catchy tune or a fashion, and to scramble these trivial items with ideas which may, as in the case of Darwinian evolution, be world-changing is a serious failure of evaluation. The endorsement by N.K. Humphrey for Dawkins's earlier book sums up both the promise and the limits of the meme:

> "'memes' should be regarded as living structures, not just metaphorically but technically. When you plant a fertile meme in my mind you literally parasitize

my brain, turning it into a vehicle for the meme's propagation in just the way that a virus may parasitize the genetic mechanism of a host cell. And this isn't just a way of talking—the meme for, say, 'belief in life after death' is actually realized physically, millions of times over, as a structure in the nervous systems of individual men the world over.'"[4]

The problem is with the use of the word 'structure.' At first it is something palpably substantial, but by the time it has re-appeared it is clearly not a structure in its full sense. It may be internalized, yet it has merely found a pool it can enter. At the heart of the matter, there seems to be what can only be called a banality not so much in the belief referred to itself, as its projected place in the human psyche.

So Daniel Dennett's highlighting of Dawkins's point that "there is no *necessary* connection between a meme's replicative power, its 'fitness' from *its* point of view, and its contribution to our fitness (by whatever standard we judge that)"[5] becomes an almost metaphysical presentation of genes. Here the failure to distinguish between an idea and a structural imprint is critical. The significance of the fourfold will be shown to be a unit crucial to human cognition and self-recognition, as well as to organic and often inorganic nature. In that sense Dennett's contention that "a human mind is itself an artifact created when memes restructure a human brain in order to make it a better habitat for memes" is a solipsism.[6] Again when Susan Blackmore in her book *The Meme Machine* writes that memes are instructions encoded in human brains as genes encode their instructions in molecules of DNA, or "in artefacts such as books, pictures, bridges or steam trains" this is a tautology, a mutually replicating definition.[7] For major books are what they are because they are built on patterns in the brain in interconnection with processes in social and natural environments.

Dawkins faces this contradiction fairly and squarely in his *The Extended Phenotype* (1982), and tries to sort out the confusion. Arguing he "was insufficiently clear about the distinction between the meme itself as replicator, on the one hand, and its 'phenotypic effects' or 'meme products' on the other," he explains:

"A meme should be regarded as a unit of information residing in the brain. It has a definite structure, realized in whatever physical medium the brain uses for storing information. If the brain stores information as a pattern of synaptic connections, a meme should in principle be visible under a microscope as a definite pattern of synaptic structure. If the brain stores information in a 'distributed' form, the meme would not be localizable on a microscope slide, but still I would want to regard it as physically residing in the brain. This is to distinguish it from its phenotypic effects [i.e. words, music, fashion styles etc.], which are its consequences in the outside world."[8]

Blackmore likewise elevates a meme to the status of "a replicator in its own right,"[9] which echoes or is analogous with the genetic replicator itself. The principal initial problem with this is that there are at least two separate definitions in circulation. Memes by the fully fleshed out definition, then, are not mere entities —a tune, the invention of a sparking plug, whatever—but a whole biological system. True, Dawkins himself approaches this when he writes of the meme as "a cultural trait [which] may have evolved in the way it has simply because it is *advantageous to itself*",[10] but there is little sense of a morphological element to his conception. Moreover Maynard Smith's has raised the crucial point that there is a lack of a proven Mendelian system for the meme's transmission.[11] So when Dennett confidently asserts such a transmission, he is basically assuming what has to be shown: "The theory of evolution by natural selection is neutral regarding the differences between memes and genes. They are just different kinds of replicators evolving in different media at different rates."[12] Elsewhere though he poses the issue more accurately:

> "But whether such [cultural] evolution is weakly or strongly analogous to, or parallel to, genetic evolution, the process that Darwinian theory explains so well, is an open question. In fact it is many open questions. At one extreme, we may imagine, it could turn out that cultural evolution recapitulates *all* the features of genetic evolution: not only are there gene analogues (memes), but there are strict analogues of phenotypes, genotypes, sexual reproduction, sexual selection, DNA, RNA, codons, allopatric speciation, demes, genomic imprinting, and so forth—the whole edifice of biological theory perfectly mirrored in the medium of culture."[13]

Certainly it is strange that a catalogue of fourfold isomorphs appear in DNA and RNA that are repeated in cultural transmission. Indeed the genetic code itself is a quaternal code made up of four symbols. DNA has the 4 bases of adenine, thymine, cytosine and guanine; RNA likewise has 4 bases, with thymine replaced by uracil. The 90% of DNA that is apparently non-functional and does not code for proteins, has a sequence of 33 sub-units repeated 4 times. This quaternal repetition is found again and again in DNA. Perhaps there is a very general connection between genetic processes and cultural activity, on the universal scale.

Dawkins at one point in his *The Selfish Gene* uses a metaphor that could strangely link these life-processes with the world of books—works of the imagination in the argument in this particular book. He writes that DNA is "written in the A, T, C, G, alphabet of the nucleotides. It is, as though, in every room of a gigantic building, there was a book-case containing the architect's plans for the entire building. The 'book-case' in a cell is called a nucleus."[14] It is interesting that evolutionary biologists talk of the *alphabet* and a *code* in relation to the construction of organisms. There is here a clue as to how defining the nature of mind, and indeed of the human condition should be approached. D.H. Lawrence's comments on myth are relevant here:

> "The images of myth are symbols. . . . They stand for units of human *feeling*, human experience. A complex of emotional experience is a symbol. . . . Many ages of accumulated experience still throb in a symbol. . . . Some images, in the course of many generations of men, become symbols, embedded in the soul and ready to start alive when touched, carried on in the human consciousness for centuries."[15]

It is the architect's plans that interest, for this skeletal fourfoldness appears so often not only in the internal world of the replicators, or genes, but in the world of objects in motion, and also in world religions and works of music and literature.

James Watson has assured me that number structure will not explain any outstanding problems in genetics (pers. comm.). And John Barrow puts forward a caveat. He warns in his *The Constants of Nature*, "there are an awful lot of numbers and even more possible permutations of them. Coincidences seem more striking because we don't think about how many unimpressive 'non-coincidences' we encounter in between finding them."[16] However, germane to our enquiry are the remarkable numerical recurrences in the *crucial* processes for life on earth which, as I will show in Chapter 2, are echoed in some of the greatest works of literature. There are certainly any series of numbers one may wish to select as significant, but in the 'skeletal' aspects of life processes the iterations stand out. I am not concerned with why this is, nor am I trying to introduce God as geometer, but at random there are many other crucial examples of this coincidence:

1. Classes of compounds essential to life are 4 in number—nucleic acids, proteins, lipids and carbohydrates.
2. The building blocks of life are four: carbon, nitrogen, oxygen, phosphorous.
3. They were all created by the simple elements of hydrogen and helium.
4. The ices of interstellar space have been shown to produce four amino acids, so that life on earth could have been seeded by meteorite action. And it is the four elements hydrogen, carbon, nitrogen and oxygen that are responsible for most of the ices in interstellar space.
5. Hemoglobin is formed from 4 amino acids.
6. As already noted, there are 4 known forces regulating the universe. Ian Stewart has observed that:

    > "Modern physics has confirmed that at its deepest levels, the universe runs on symmetric lines. (*In maths, an object shows symmetry if it maintains its form after some transformation*). Principles of symmetry govern

the four forces of nature (gravity, electro-magnetism, and the strong and weak nuclear forces that act between fundamental particles); the quantum mechanics of elementary particles; the nature of space, time, matter, and radiation; and the form, origin and ultimate destiny of the universe. We don't know *why*, but we are pretty sure that it is so."[17]

7. In the genetically archetypal fly *Drosophila melanogaster*, there are 4 pairs of chromosomes, and at each cell division during the development of the egg into the adult, the chromosomes are reproduced so that each cell in the adult body resembles the fertilized egg in having two similar sets of 4 chromosomes.

8. The cerebrum of the human brain has 4 paired and major lobes of its own, and under this forebrain the remainder is similarly of 4 parts.

9. There are some odd 4% and 0.4 of 1% figures for changes in electro-magnetic forces.

    Then again, Lynn Margulis has distinguished four once entirely independent and physically separate bacterial ancestors that merged to become the green algal cell. The process of symbiogenesis involved four steps:

    The ground substance of cells, the nucleocytoplasm is descended from archaebacteria.

    Cilia, sperm tails, sensory protrusions arose in the original fusion of the archaebacterium.

    The oxygen-respiring mitochondria in our cells and other nucleated cells evolved from bacterial symbionts now called 'purple bacteria' or protobacteria.

    The chloroplasts and other plastids of algae and plants were once free-living photosynthetic cyanobacteria .[18]

10. The insect body: each segment of an insect consists of a firm chitinous hoop built up of 4 parts: a large dorsal plate (Tergum), a large ventral one (Sternum); they are united by a soft region (Pleuron). The segments are joined in from and behind by a soft fold which allows movement, and each bears a pair of jointed walking limbs.

    This segmentation is not merely external, but also affects the internal organs, so that the segments have their own muscles and nerve-supply.

11. (The body is sub-divided into 3 more regions: head, thorax or chest and abdomen. All insects possess 3 pairs of true jointed legs, and these are developed on the 3 segments of which the thorax is invariably composed, one pair to each.)

12. The mouth-parts of a biting insect (beetle or cockroach) are four in number.

13. Ecdysis, or moulting, takes place (generally) four times in butterfly larvae.

14. Evolution has at least four ingredients: mutation, selection, development, environment. There have been 4 major stages in the evolution of *Homo sapiens*:
    i) various *Australopithecus*
    ii) *Homo habilis*
    iii) *Homo erectus*
    iv) *Homo sapiens*
15. The first chapter of *Promethean Fire: Reflections on the Origin of Mind* by Charles J. Lumsden and E.O. Wilson is entitled 'The Fourth Step of Evolution:' "The turning point at which such a fully human head and human mind began to be fashioned upon the already human body can be regarded as the latest of the four great steps in the history of life on Earth. These events, which occurred roughly one billion years apart were first the beginning of life itself in the form of crude replicating microorganisms; then the origin of the complex (eukaryotic) cell through the assembly of a nucleus, mitochondria, and other organelles into a tightly organized unit forming the basis of all higher life; next the evolution of large, multicellular organisms (flatworms, crustaceans), which could evolve complex organs such as eyes and brains; and finally the beginnings of the human mind."[19]
16. Derek Bickerton distinguishes 4 stages in human social development:
    i) 200–40 kya—hunters and gatherers - blade technology and language which took place in tropical and subtropical regions.
    ii) 40–10 kya—some went north, beat the Neanderthal, and became king of the steppes, negotiating a less friendly environment. They Domesticated plants and animals which triggered—
    iii) 10kya–400 years ago—from loosely organized nonterritorial creatures changed to tightly controlled territorial. Technology, inequality, violence became characteristic in their new, post-agricultural organization.
    iv) Powers expand, numbers increase, specialization of labour, and wealth increases.[20]

However I am not arguing that cultural evolution can be seen as a direct reflex of fundamental physical fours. Rather I am putting forward the hypothesis that the plethora of significant fours that provide a structural backbone to many, if not most, of human culture's greatest achievements are variations on a type of Cosmic Cultural Faculty, a structure imprinted in the human brain and frequently appearing in key variations in the cosmos. As Ian Stewart puts the issue, though he would posit an answer to the problem by a very different route:

"Senses produce internal patterns of neural activity in brains; in order to be useful, those patterns must correspond, in some manner, to significant patterns in the outside world. Therefore, the neural nets used in sensory perception must be organized in a way that reflects the deep patterns of the external universe . . . so it is only to be expected that the symmetries of the universe should somehow be imprinted on our sensory apparatus."[21]

There is an innately specified set of structures, without any of the simple switches to turn it on that appear in Chomsky's grammar. The concept of 'meme' can be a useful way of referring to this structure, and much of the evidence I shall adduce will approximate to and have something of the disparateness that Dawkins and Blackmore ascribe to memetics. But my overall contention will be that religions and great works of music and literature are fusing a number of faculties in the brain and senses. It is not so much that these structures are advantageous to themselves in the way that genetic replication is, but that in their relation with non-human nature they are part of the human condition, part of being at-home in nature and the cosmos. The symmetries are also integrated into an implicit message in language that will be excavated. They are autonomous in the way that Blackmore describes memes, but it makes no sense to call them "selfish."[22] That is simply an aspect of the idealization of Genetics by the memeticists. But they do have something of the character of an algorithm. And as two writers who focus on the symbiotic nature of the evolution of species point out "selfish genes, since they are not selves in any coherent sense, may be taken as figments of an over-active, primarily English-speaking imagination. The living cell is the true self."[23] That is, they are part of a process which is both simple and random, but in which the cell is the unit of life. The processes of the Cosmic Cultural Faculty are closer to the description of the order of creative structures in T.S. Eliot's "Tradition and the Individual Talent":

> "The historical sense compels a man to write not merely with his own generation in his bones, but with a feeling that the whole of the literature of Europe from Homer and within it the whole literature of his own country has a simultaneous existence and composes a simultaneous order. . . .
>
> The existing monuments form an ideal order among themselves, which is modified by the introduction of the new (the really new) work of art among them. The existing order is complete before the new work arrives; for order to persist after the supervention of novelty, the *whole* existing order must be, if ever so slightly, altered; and so the relations, proportions, values of each work of art toward the whole are readjusted; and this is conformity between the old and the new."[24]

Elsewhere, Eliot's description of the casting off of personality in artistic creativity and facilitation of a greater defining force is very like variation of genetic material during replication, where the poet or artist is the vehicle for the transmission of the fourfold backbone:

"What happens is a continual surrender of himself as he is at the moment to something which is more valuable. The progress of an artist is a continual self-sacrifice, a continual extinction of personality."[25]

It is significant that Eliot's final great work was *The Four Quartets*.

NOTES

1. Dennett (2), 347.
2. Dawkins (4), 323.
3. Dawkins (4), 192.
4. Dawkins (4), 192
5. Dennett (2), 363.
6. Dennett (2), 365.
7. Blackmore, 17.
8. Dawkins (2), 109.
9. Blackmore, 30.
10. Dawkins (4), 201.
11. Smith, 47.
12. Dennett (3), 128.
13. Dennett (2), 345.
14. Dawkins (4), 22.
15. Lawrence (1), 49.
16. Barrow, 72.
17. Stewart (1), 38.
18. Margulis, *passim*.
19. Lumsden (2), 7.
20. Bickerton, 244.
21. Stewart (1), 165–66.
22. Blackmore, 8.
23. Margulis and Sagan, xvi.
24. Eliot (1), 14–15.
25. Eliot (1), 17.

Chapter Three

Shakespeare and the Nature of Verbal and Musical Language

Shakespeare tends to write in segments of 4 plays. So the peak of his early work is the two historical tetralogies. Later there are the 4 great Tragedies where his heroes and heroines meet their fate in 5-Act worlds. These are interwoven with 4 Problem Plays followed by the 4 Last Plays. Keats's much quoted perception of Shakespeare's biographic parabola can be seen anew in the light of these paradigms: "A Man's life of any worth is a continual allegory—and very few eyes can see the Mystery of his life—a life like the scriptures figurative—which such people can no more make out than they can the hebrew Bible. Lord Byron cuts a figure—but he is not figurative—Shakespeare led a life of Allegory; his works are the comments on it."[1]

Working from this analogical method, the following reveals itself:

1. The Egg (the History Tetralogies—3 Parts of *Henry VI*, *Richard III*, and then *Richard II*, 2 Parts of *Henry IV*, and *Henry V*)
2. The Larva (The Problem Plays—*Much Ado About Nothing*, *Troilus and Cressida*, *All's Well that Ends Well*, *Measure for Measure*)
3. The Pupa (The Tragedies—*Hamlet*, *Othello*, *Macbeth*, *King Lear*)
4. The Imago (The Last Plays—*Pericles*, *Cymbeline*, *The Winter's Tale*, *The Tempest*)

In Shakespeare, the creative process can be seen as a fourfold helix associated with the parallel processes of metamorphic evolution. For Henry James, the culmination and cessation of Shakespeare's artistic career in *The Tempest* is a mystery whose "power to torment us intellectually seems scarcely to be borne:"

"What manner of human being was it who *could* so, at a given moment, announce his intention of capping his divine flame with a twopenny extinguisher, and who then, the announcement made, could serenely succeed in carrying it out?. . . . [It] puts into a nutshell the eternal mystery, the most insoluble that ever was, the complete rupture, for our understanding, between the Poet and the Man."[2]

But Shakespeare clearly sensed he had completed his series of fourfold helices. It was not a conscious thing, but so perfectly natural and spontaneous was Shakespeare's art that it had reached a perfect conclusion over the vast canvas of his overall development. Schelling grasped this when he wrote that "Shakespeare never portrays either an ideal or a formal world but always the *real* world." But when he goes on to say that "the ideal element manifests itself in the construction of his plays," Schelling has turned the world upside down positing the ideal for the real, since the construction unconsciously conceals an entirely natural and material fourfold significance.[3] As James implies, Shakespeare the ordinary citizen then wanted release from the exigencies of his art:

"Here at last the artist is, comparatively speaking, so generalized, so consummate and typical, so frankly amused with himself, that is with his art, with his powers, with his theme, that it is as if he came to meet us more than half-way, and as if, thereby, in meeting *him*, we were nearer to meeting and touching the man."

This was artistic resignation, or death, the dramatic poet earthed. Henry James's hand on his deathbed was still making the motions of writing with a pen. But Shakespeare had already achieved what James and the novelists continued to strive for, "some copious equivalent of thought for every grain of the grossness of reality . . . the joy of sovereign *science*."[4]

George Steiner's pontification is quite erroneous when he announces that "Shakespeare's manifold and secular humanity is unreceptive of systematic unification," which echoes Schelling's criticism that "he is too diffuse in his universality."[5] The spontaneity and naturalness of the dramatist seems to annoy the critical theorists. Again Steiner writes, somewhat irritably, "In most writers, Shakespeare representative among them, the compositional process seems to show no correlations with what we know of the methods of discovery in mathematics. But in some poets (Poe or Valéry, for example), as in musicians, painters or architects, the affinity to mathematical means and ideals is significant. They feel, they construct *more geometrico*."[6] The basis of medieval psychology has already broken up by the time of Shakespeare, the structure of the four humours set upon the four elements absorbed, but the whole is transformed and extended in relation to personality almost beyond recognition. So, for example, Ben Jonson's plays, despite all their dexterity and wit, plod in comparison, tied to an already surpassed concept of temperament. Shakespeare's plays embody in their full span a scientific apprehension of the creative mind grounded in an unrivalled understanding of human motive and action. What remains is to link it with the universe.

Heidegger infamously remarked of the fullest body of knowledge about the world around us: "science does not think." Much of science merely records, however elaborate the theoretical template required or complex the experimentation. Imagination is not so directly bounded by empirical fact. It lingers on the edge of consciousness, only activating when it can hitch a ride on passing fragments of intuition and go on to assemble newfound connections into an entirely new, scientific or poetic, form. Newton famously transmuted the falling apple in his Lincolnshire garden into knowledge of the gravitation forces applying through space and thus to the orbit of the moon, and this is clearly an act of imagination followed up by a formidable analytical intelligence. Then again, centuries later, Einstein in a staggering feat of imaginative reassemblage firstly conceived a constant velocity for light, and then reassessed the whole of contemporary physics. But whether even such an application is at the highest level of imaginative activity will be the question this essay addresses. Imagination is connected to a primal human impulse just below the threshold of consciousness, and at its highest philosophical and poetic regions relates to inner reality rather as Minkowski's four-dimensional mathematics relate more exactly to the external world than immediately accessible and commonsensical three-dimensional concepts.

This does not prevent some scientists from rejecting the role of art, just as some philosophers and critics downgrade science. So Paul Dirac believed what he called the "pretty mathematics" lying behind reality rendered the poetic imagination obsolete:

> "I do not see how a man can work at the frontier of physics and write poetry at the same time. They are in opposition. In science you want to say something nobody knew before, in words which everyone can understand. In poetry you are bound to say something that everybody knows already in words that nobody can understand."[7]

Henry David Thoreau had put the counter argument a century earlier: "I should say that the useful results of science had accumulated, but that there had been no accumulation of knowledge, strictly speaking, for posterity; for knowledge is to be acquired only by a corresponding experience."[8] This mid-nineteenth century statement, though, has been comprehensively rendered obsolete by twentieth century science. Einstein put the matter plainly in relation to modern gravitation theory: "no ever so inclusive collection of empirical facts can ever lead to the setting up of such complicated equations. A theory can be tested by experience, but there is no way from experience to the setting up of a theory."[9] However even though the knowledge available through scientific discovery can lead to changes in the behaviour of organisms, there remains a question as to whether in the last resort this is any more than a higher form of information.

Even if this is the case, poetry has no cause for complacency. Among contemporary poets only Miroslav Holub has successfully poetically engaged with science, though earlier Paul Valéry had characters debate Einstein in his *L'Idée Fixe*. A nineteenth century poet has recently been shown not merely to have foreseen Darwinian natural selection, but to have incorporated into his poetry, structures of thermodynamics only in the process of being discovered by Lord Kelvin and others. I mean Tennyson, and his epic elegy, *In Memoriam*, published in 1850 but gathering together poems written over a period of 17 years. It is well known that the sections of this poem regarding Evolution had been circulating for several years prior to Robert Chambers's limited but pioneering *Vestiges of the Natural History of Creation* of 1844. What has been less well appreciated is that, as a recent article by Barri J. Gold reveals, "*In Memoriam* is saturated with the language of energy physics," so much so that she concludes that Tennyson "can be said to have discovered—poetically—not only the terms but also the principles and processes of the nascent science of energy physics, especially the poetic evocation of the tension between conservation and dissipation that haunts the first and second laws of thermodynamics."[10] In other words, a truly great poet can intuit advances in science, and indeed scientists of the mid-nineteenth century consulted Tennyson in regard to their latest work. A parallel situation for a poet is almost inconceivable in the Western world today, yet poetry will not reclaim the intellectual and cultural high ground beyond the national and ethnic into which it has stumbled until it recurs.

It is significant that Miroslav Holub has endorsed experimental psychology's conclusion that "the present moment lasts three seconds."[11] This is a piece of knowledge that is provocative to the imagination. For if experience and reaction to *immediate* events appertain to the triple, then we might expect reflection, creativity and religious systems together with the natural systems to which they unconsciously refer to be grounded in the fourfold. Etymology in relation to entomology provokes an interesting conundrum in regard to the issue at stake, and may disprove Heidegger's contention that "in order to be who we are, we human beings remain committed to and within the essence of language, and can never step out of it and look at it from somewhere else."[12] Language can propose a higher metamorphosis. It does of course provide a selective advantage to the human species, and as such has a dialectical relationship with genes and environment. But it also offers key words carrying a 'message' that require interpretation. So William Blake's Los is eternally reconstructing language for its inner meanings, empowering symbols to enable the individual to view all four sides simultaneously, imaginatively.

One of the associative components of the word 'imagination' is imago — an image, or more to the point here, the inclosed image in the caterpillar that comes forth as the adult of an insect. If the associations of imago are traced back through language, then Latin 'pupa' — the root of pupil both as student and eye — is the chrysalitic stage of insect development, with something of the character of krusos, gold, in the root of the associated process, the 'chrysalis'. Pupa also relates to the pupil of the eye, the twinkling of a thought or theory long before the full work of realized imagination. As Steiner records, "the comparison of the pupil of the eye to a small child (*pupilla*), has been traced in all Indo-European languages, but also in Swahili, Lapp, Chinese, and Samoan."[13] Tunneling further down language takes us to a fundamental word, 'larva', which signifies in the Latin both person and mask. So when Descartes writes "larvatus prodeo"[14] — "I proceed masked" — the condition of the insect becomes a metonymy for a certain stage of the human condition, as observations may prepare the way for theory through the use of applied imagination. Einstein's remark to Heisenberg sent the latter on the trail of the Uncertainty Principle: "It is the theory which decides what we can observe." The caterpillar masks the perfect imago, embryonically contains its enzymes and cells, so likewise the radix of the person is the Latin persona, or actor's mask. By way of these etymological roots, the imagination becomes the true measure of the growth of the conscious human, while at the same time it relates us to the entomological, by way of the succession ovum-larva-pupa-imago. The human species has evolved from the great apes, but the inner labyrinth of language connects us to the insects.[15]

I am proposing that this underground river of meaning ultimately creates a four-dimensional ontology for language and the arts, isomorphic to the four-dimensional ontology of perduring objects favoured by the special theory of relativity. Both go beyond inertia systems. Neither are immediately accessible to referential understanding because of the dynamics of shifting co-ordinates. The totality of the course of a symphony is to be understood as an abstract formulation like Minkowski's concept, a mathematical analogy breaking up the fixity of place time. The next chapter shows how the music of Beethoven, Mozart, Brahms or Bruckner disperses our everyday self, and then dramatizes in musical notation and structure a metamorphic progression which it coaxes out from behind the unconscious and mask of that self. Worldlines or worldworms of personal development become, for the extent of the symphony or quartet, akin to string bundles of psychic events stretched out in the space-time of the unconscious. Through the great works of the imagination, the individual travels up and down the passages of self, undergoing reversion and regression as well as progression.

Holub's emphasis on the significance of the three-second moment can now be recognized for the brilliant insight it is. It establishes the limits of the threefold in the world of the intellect, belonging to what Valéry called "moi no. 2," or the adaptive self acting under the aegis of experience. The imagination, though, can operate in the quaternal sphere and is a way of identifying the components that can lead to new knowledge. It does this by discovering similarities between things whose likeness had not previously been realized in thought. Indeed the great works of imagination in music and literature can illuminate the "now" that so exercised and worried Einstein. This is because these works move through time, and yet being abstracted from real time, reveal things that are apparently beyond time. The structures they are built on relate directly to the reader/observer who is the ultimate co-ordinate of "now." Whereas Relativity theory places all time on one curve so that past, present and future are contemporaneous, Quantum physics makes time entirely uncertain, to the degree that an atom does not know its next move. Indeed the superposition principle enables a particle to be in two places simultaneously. But neither Relativity nor Quantum mechanics can explain the human, subjective, sense of the flow of time. Einstein recognized this when he asserted "there is something essential about the 'now' which is outside the realm of science," the experience of which has a special significance for humans. The four-dimensional continuum results in 'now' forfeiting its objectivity in the spatially extended world, and so is left with only a 'frame-relative' existence.[16] Then Gödel goes on to define reality as consisting of "an infinity of layers of the 'now' which come into existence successively."[17]

A scientific treatise demands a point-by-point understanding of the stages of the proof or argument, but the works of the imagination immediately penetrate a core of intellect even before what lies behind them can be unraveled. They transform the inner life of an individual by a type of ghostly, abstract experience which is the work of art. Ultimately imagination is an intellectual tool, whereas knowledge is a product of the use of that tool. Knowledge is what the imagination sloughs off in its own capacity as what Nabokov calls "the muscle of the soul."[18]

NOTES

1. Keats, 2: 67.
2. James, 438. James's emphasis.
3. Schelling (3), 271. With modifications in translation.
4. James, 432, 431.
5. David Simpson, 137.
6. Steiner (3), 73–74.
7. Farmelo, *passim*.
8. Thoreau (5), 364–65.
9. Schlipp 1:89.
10. Gold, 450–51.
11. Holub, 1.
12. Heidegger (1), 134.
13. Steiner (1), 102.
14. Descartes, 213.
15. Norman O. Brown's epic of associative thought, *Love's Body*, suggested part of this linguistic structure (Brown, 96–97).
16. Einstein (1), 149.
17. qd. By Yourgrau, 164.
18. Nabokov (12), 77.

Chapter Four

Schopenhauer, Music
and the Order of Things

The World as Will and Representation locates the uniqueness of music in its capacity to replicate the sense of willing. On the other hand, Mallarmé valued music not for its euphonic elements, but for its structure. Putting these two definitions together suggests that the human will is predicated on a purposive search for symmetry, that in responding to music the human brain is not seeking out the superficial elements of sound, though it may be these that lead the listener into the labyrinth. When threaded into music, the will intimates some crucial element of human life itself. Music in the form of song associated with work rhythms probably preceded verbal language and the literary arts of lyric and tragic verse as Nietzsche suggests in *The Birth of Tragedy from the Spirit of Music*, so that it appeals to some primary life-impulse. In becoming engrossed in music until a sense of everyday time is lost, we enter the heartland of what I have defined as the Cosmic Cultural Faculty, a faculty which links us purely directly, if intuitively, to the patterns of the cosmos. As one leading researcher into the musical brain has vividly put it—"When music causes one of these 'skin orgasms,' the self-reward mechanisms of the limbic system— the brain's emotional core—are active, as is the case when experiencing sexual arousal, eating or taking cocaine."[1] Although there have been various attempts to lump this experience in with the main demands of natural selection, and thus discover a utilitarian and social function, this is merely routine sociological thinking.

Steven Pinker goes so far as to reduce the issue to the question: "if music confers no survival advantage, where does it come from and why does it work? I suspect that music is auditory cheesecake, an exquisite confection crafted to tackle the sensitive spots of at least six of our mental faculties."[2] But this is too simple and arises from his inability to estimate the impact of the shape of music, its patterning and appeal to human sensibilities. So he writes that "music communicates nothing but formless emotion."[3] Actually music expresses four structures; melody, rhythm together with metre, timbre alongside tone, and volume plus dynamic progression. The brain works through a considerable series of processes in order to organize this variegated set of events, and the more experienced the listener or practitioner, the faster these are absorbed. There are specific neural circuits that filter the music. Eckhart O. Altenmüller describes the system:

> "After sound is registered in the ear, the auditory nerve transmits the data to the brain stem. There the information passes through at least four switching stations . . . [after which] the thalamus—a structure in the brain that is often referred to as the gateway to the cerebral cortex—either directs information on to the cortex or suppresses it. . . . Early stages of music perception, such as pitch (a note's frequency) and volume, occur in the primary and secondary auditory cortices in both hemispheres. The secondary auditory areas, which lie in a half-circle formation around the primary auditory cortices, process more complex music patterns of harmony, melody and rhythm (the durations of a series of notes). Adjoining tertiary auditory areas are thought to integrate these patterns into an overall perception of music."[4]

Schopenhauer wrote that "*metaphysics is impossible* as being the science of that which lies beyond nature, that is, beyond the possibility of experience."[5] Nonetheless experience itself is ambiguous. Insect metamorphoses seem to be remote from human experience. Even in artistic works, they seem alien except for works such as Kafka's *Metamorphosis*, Hardy's *The Return of the Native,* or David Cronenberg's film 'The Fly.'

But if we look at classical music, matters are not so simple. Take the usual structures of symphonies and quartets. A symphony is a sonata for orchestra with, normally, four movements. In the first movement, themes are stated (the egg, or by some recent interpretations the proto-larva); the second movement proceeds slowly like a caterpillar; the third such as the scherzo of the *Eroica* tends to be febrile and anticipatory like a shimmering chrysalis trembling with incipient finality; while the fourth usually represents a summation, which as Berlioz analyzed in relation to Beethoven's composition, leads "from tension to release, from compulsion to liberation, from the tragic to the joyous;"[6]—with exceptions like the tragic last movement of Brahms's Fourth Symphony. Undercurrents of profound sadness often underlie the "most capricious evolutions" of a Beethoven scherzo, and I would interpret this as the process of the loss of imaginal buds which accompany the pupal or chrysalitic stage of development, essential to achieve the state of the imago, but nonetheless a loss of youthfulness and an intimation of the end. As lis-

teners or performers, we travel through these stages as the musical form un-
folds. The classical style constructed itself on the four-measure phrase once it
had broken from the flowing continuity of the Baroque. The great Cantatas
had decorated their music with vocal and instrumental colour like the painted
church images. As Spengler puts it: "Music frees itself from the bodiliness in-
herent in the human voice and becomes absolute. The theme is no longer an
image but a pregnant *function*, existent only in and by its own evolution, for
the fugal style as Bach practised it can only be regarded as a ceaseless process
of differentiation and integration"[7] So music became more abstract in the late
eighteenth and nineteenth centuries. Charles Rosen estimates that it was
about 1820 that the four-measure unit gained pre-eminence in rhythmic structure,[8] but the four-movement symphony had already become dominant even
by 1780, although three-movement ones continued to be composed. In other
words, the very fabric of the great period of classical achievement is both an
anthem to quaternity, and a reverberation of the insect connection. Robert
Simpson catches something of this contradiction when he writes that the last
movement of Beethoven's Ninth symphony has the composer saying in effect
"the visions of the first three movements are such as to reduce man to the apparent size of a microbe; but a man conceived them, so let us all rejoice in our
potentialities."[9] The great works of Haydn, Mozart and Beethoven, and those
who followed them, are metaphysical to our consciousness, yet have a very
real basis in the natural world. Our experience of them is at one and the same
time seemingly immediate, and yet so ghostly and phantom-like.

More than this, though, there are meta-musical patterns also. Brahms wrote 4
symphonies, as did Schumann, and his final works include the 4 Piano Pieces
and 4 Serious Songs. So also of course Richard Strauss's 4 Last Songs. Brahms's
Second Piano Concerto does not follow the usual 3-fold concerto shape, but rises
to the symphonic four movements. He originally intended his Violin Concerto to
have four movements, while his first Piano Concerto was conceived as a symphony based on Beethoven's Choral Symphony. All Brahms's symphonies have
a four-movement structure, and all of their first movements follow a sonata
shape. Their totality constitutes a hymn to the Tetradic, with their particular thematic continuity and integration through the four sections. Schubert also, who so
sought to emulate Beethoven, stretched out towards a series of fourfold works,
his compositions being "the product of my mind and spring from my sorrow;
only those that are born of grief give the greatest delight to the outside world."[10]
Three crucial works though are left unfinished: the Quartettsatz in C major, and
the E major and 8th B minor Symphonies. As Wilfrid Mellers explains, Schubert

"finally solved his most difficult technical and imaginative problem. He had
resolved drama into song; and in the andante of the B minor had followed this

resolution with the bliss of Eden. He could not rest permanently in a recovered Eden; but at this stage in his career he could not see how he could continue without descent or bathos. He found an answer only in the last three quartets and the C major Quintet, composed during the last four years of his life."[11]

In Blakean terms, Schubert hovered here around the threefold gates of Beulah, while his last four sonatas are the apogee of his writing for piano. Later Bruckner would build the bar cells of his symphonies using the quaternities of the Mass as his basis. So the final fourth movement of his 5th Symphony unifies the themes of earlier sections in "a double fugue which also embraces elements of sonata style ," which also echoes the Agnus Dei of his F minor Mass.[12] Schumann though sensed that the first part of the nineteenth century was the apotheosis of the sonata-based symphony, and wrote in 1835 that he "almost feared that the term 'symphony' might soon become a thing of the past," while by 1839 he feared that "isolated beautiful examples of [the sonata form] will certainly still be written now and then—and have been written already—but it seems that this form has run its life course."[13] (Interestingly these dates approximate to the epoch of the last indisputably great English poetry running from Blake to Shelley and Coleridge, after which Tennyson began to absorb the new sciences.) Although Michael Tippett kicked against the historical archetype of the classical symphony in favour of the notional archetype which permits endless variation, even he ended up writing four symphonies.

The failure of the concept of 'meme' is clearly exemplified in the comments of its advocates on Beethoven's Fifth Symphony. Daniel Dennett surmises that:

> "The first four notes of Beethoven's Fifth Symphony are clearly a meme, replicating all by themselves, detached from the rest of the symphony, but keeping intact a certain identity of effect (a phenotypic effect), and hence thriving in contexts in which Beethoven and his works are unknown."[14]

The overall structure of the symphony follows the general pattern of the fourfold quartet movement, the slower third movement running into the finale without a break. The first four notes of the piece are therefore an urgent microcosm of the whole symphony, not only the thematic basis of much of it, but a stamping out or imprinting of the total mathematical impact. In Beethoven's words, they are "destiny knocking at the door," and these four raps also appear in the Appassionata, the G major Piano Concerto and the 3rd movement of the *op*. 74 Quartet. It is especially clear that to take these initial notes as a catchy meme is seriously inadequate. They are part of the groundswell of the (re-) awakening of a biological faculty, an empowering and invigorating of our capacity to structure and unify thought and emotions. Most immediately the four-note motif is based on the song of the yellow-hammer, according to Beethoven's pupil, Carl Czerny. The Scribbling Master's birdsong is reversed in the first four notes of the Symphony, but optimistically turned the right way round ending on an upward note as it moves into the final movement. Blackmore follows Dennett into memeland: "Is Beethoven's *Fifth Symphony* a meme, or only the first four notes? . . . Whether by coincidence or by memetic transmission, Beethoven is the favourite example for illustrating this problem."[15] She goes on to refer to the opening ta-ta-ta-tum as the "good old four notes" in the same context as "any catchy tune,"[16] demonstrating yet again the total failure of symmetrical thinking in memetics. Robert Layton's analysis of the apotheosis of the four taps transformed in the last movement gives a clear idea of the gulf between the meme machine, and what is actually going on in Beethoven's Fifth. He describes the journey undergone in the symphony as it makes the transition to the Finale which "is in every respect a momentous event, whether we see it as the acting out of a revolutionary age's will to transform; or as Beethoven's own spiritual ascent from within, the moment when the composer, increasingly deaf and isolated, moves from the gloom of the mind's inner landscapes to greet with joy a daylight world of heroic action."[17]

Indeed the mention of "a daylight world of heroic action" suggests all that is wrong with the concept of selfish gene. Of course if one interprets any altruistic act as merely self-interested even where it involves a sacrifice for humanity as a whole, then action in life is always such an expression. How, though, do we explain the volunteering to the International Brigades in 1936 when people from multiple countries went against government indifference to put their lives on the line to fight Fascism? Little wonder that the anti-Mussolini Arturo Toscanini is one of the great interpreters of Beethoven's Fifth Symphony, even if at Queen's Hall, London in 1937 he insisted when conducting the *Eroica* that "Is-a not Napoleon! Is-a not 'Itler! Is-a not Mussolini! Is-a *Allegro con brio*!" Although the meme idea is initially provocative and apparently plausible, its essential triviality is painfully exposed when referring to anything greater than, say, "Dancing in the dark."

Schopenhauer's description of music makes it clear that for him Leibniz's arithmetic of music is only the husk of the issue, a clue to the profounder reverberations. Because it appeals so forcefully to humanity's innermost being through a universal language, it goes beyond even the world of perception. And similarly it transcends Leibniz's "unconscious exercise in arithmetic," because while that is fine as far as it goes it only touches the exteriority of the matter and only provides the satisfaction of a sum that comes out right. In fact music appeals to and expresses "the deepest recesses of our nature."[18] Schopenhauer attributes to music a key role in what was after his death to be defined as natural selection. Music is seen as crucial to the will to survive of the individual since it is a *"copy of the will itself*, the objectivity of which are the Ideas. For this reason the effect of music is so very much more powerful and penetrating than is that of the other arts, for these others speak only of the shadow, but music of the essence."[19] Schopenhauer is offering a purely abstract description which yet makes precise knowledge of the elusive 'thing-in-itself' that Kant would not admit. Absorbing oneself in a symphony allows the individual to see the whole of her or his life passing along, for "music differs from all the other arts by the fact that it is not a copy of the phenomenon, or, more exactly, of the will's adequate objectivity, but is directly a copy of the will itself, and therefore expresses the metaphysical to everything physical in the world, the thing-in-itself to every phenomenon."[20] Music is a familiar yet remote paradise which is perfectly accessible to us, and yet, as Schopenhauer sees it, fundamentally different from our true nature and our environment. What it expresses are "the innermost stirrings of our will, that is to say of our true nature," and since we are not usually aware of these aspects of the essential self, music seems strange and remote."[21]

The strangeness and remoteness of music though arises because it shadows the world of the insects in their metamorphic evolutions, and that unbeknown to us it transports us to that world which paradoxically is also our own internal world of self-evolution. This has a cosmological element as the experience echoes in a ghostly manner contemporary humanity's relationship to the fundamentals of relativity. The listener or practitioner of music is, mentally speaking, in motion as in Aloys Wenzl's description of the twentieth century scientist:

"One could say, that the observers of moving systems are like Leibniz's monads and that Leibniz's idea of a pre-established harmony finds an analogy in the theory of relativity. Just as the world is mirrored differently in each monad and yet the sights of all monads are related to each other and translatable into each other, so also does the 'absolute' four-dimensional world-continuum appear in different values of spatial and temporal measurements to every observer imprisoned [as he is] in his own system, yet all sights are transformable into each other."[22]

Schopenhauer is necessarily working within the postulates of Newton's absolute space and absolute time. In that sense he is shackled by the limits of the intellectual spirit of the age. So when he comes to the more oblique questions, he is drawn into making mystical projections of consciousness: "There is something which lies beyond consciousness but which sometimes breaks into this, like a moon-beam into a clouded and overcast night. . . . It is our essence-in-itself that lies outside time."[23] Music is the clue to the inner nature of the phenomenal world missed by his great progenitor, Kant, and is "so completely and profoundly understood by him in his innermost being as an entirely universal language, whose distinctness surpasses even that of the world of perception itself." It is "as it were the innermost soul of the phenomenon without the body."[24]

Bryan Magee in his book on Schopenhauer argues that

"the reason why words cannot dig down to the same level [as music] lies in their excessive generality. . . . Language cannot make use of concepts which are formed by a process of generalization, and this means that it can never communicate insight into the unique in-itself-ness of anything. Music, however, does. And in doing so it is 'completely and profoundly understood by [man] in his innermost being as an entirely universal language whose distinctiveness surpasses even that of the world of perception itself.' Now since what a philosopher like Schopenhauer is trying to do is to formulate an 'expression of the inner nature of the world in very general concepts', it follows that the composer is already doing *in concreto* what the philosopher is attempting to do *in abstractio*. Therefore if, *per impossibile*, we could succeed in giving a 'perfectly accurate and complete explanation of music which goes into detail, and thus a detailed rehearsal in concepts of what music expresses, this would also be at the same time an adequate rehearsal and explanation of the world in concepts, or one wholly corresponding thereto, and hence the true philosophy. . . . The composer reveals the innermost nature of the world, and expresses the profoundest wisdom, in a language that his reasoning faculty does not understand.' His is therefore the purest, the most undiluted form of genius of all, because it is the least contaminated by conceptual thought or conscious intention."[25]

It is precisely the *"per impossibile"* that is the problem. This is the burning core of the intellectual volcano that contains many unanswered questions and crucial knowledge links; in other words what knowledge of the noumenon depends on, and where it can be found. Get to grips with this, and there will be an extension to the theory of evolution, distant though for the moment the mainstream scientific world appears from this issue. We have to go to the roots of key words and to a helical series of words—to *paroles* rather than *langage* in Saussure's and Barthes's terms—to cast further light on insect-human inner relations. But for this I shall have to shift the focus and move from philosophy to literature itself, moving from a writer close to the processes of metamorphosis—Apuleius—to a revelation of word-clusters that relate to the structures of music under discussion. It will also take in the mutations of identity in some of the great works of literature, since these mutations are the human parallel of transformations in the insect world.

NOTES

1. Altenmüller, 21.
2. Pinker, 534

3. Pinker, 529.
4. Altenmüller, 26–28.
5. Schopenhauer (3), 1:81.
6. Berlioz, 66.
7. Spengler, 1:282–83.
8. Charles Rosen, 58.
9. Robert Simpson, 57.
10. quoted, Mellers, 3:91.
11. Mellers, 3:92.
12. Mellers, 3:112.
13. quoted, Ballantine, 32.
14. Dennett (2), 344.
15. Blackmore, 53.
16. Blackmore, 56.
17. Layton, 95.
18. Schopenhauer (5), 1:256.
19. Schopenhauer (5), 1:257.
20. Schopenhauer (5), 1:262.
21. Schopenhauer (1), 3;11–12.
22. Wenzl, 586.
23. Schopenhauer (1), 3:629–30.
24. Schopenhauer (5), 1:256; 2:262.
25. Magee (2), 183–84.

Chapter Five

From Apuleius to A.R. Wallace: Evolutionary Theory and Some Literary Animals and Insects

Reitzenstein has made a useful distinction between *fabula* and *historia*, where the latter is a cultivated and consciously literary product and the *fabula* an oral tale, or imitation of one.[1] However it is a feature of the finest contemporary writing that it treats the distinction with the greatest flexibility. So Gabriel García Márquez frequently fuses the two. His *A Very Old Man with Enormous Wings* relates the frustration of a rural community at the arrival of a perhaps-angel, a perhaps-vulture of death, or simply a perhaps-fraud. But at the end when the old man's wings have recovered sufficiently for him to leave the village, he is apprehensively watched away by a peasant who is relieved "he was no longer an annoyance in her life but an imaginary dot on the horizon of the sea."[2] The community prefers its own stoic ethic to a saviour or outside *mores*. Márquez's Macondo has succeeded Faulkner's Yoknapatawpha in directing one branch of contemporary fiction towards what Salman Rushdie has defined as the elevation of "the world-view above the urban one; this is the source of his fabulism."[3]

Approaching this from the diametrically opposite perspective, George Eliot's Casaubon was thoroughly, and damagingly, contemptuous of "the fable of Cupid and Psyche, which is probably the romantic invention of a literary period, and cannot, I think, be reckoned as a genuine mythical product."[4] Such an attitude to created fable is not uncommon outside fiction also, and a disabling genre purity is characteristic of much Western European criticism. But William Blake had already grasped the achievement of Apuleius:

"Apuleius's Golden Ass & Ovid's Metamorphoses & others of like kind are Fable; yet they contain Vision in a sublime degree, being derived from real Vision in More ancient Writings."[5]

And later Coleridge looked on *The Golden Ass* as "a sort of hybrid Poetry —Like the rough Copies of Hints taken down by a Poet," a hybrid writing which overthrew the rigidity of informative genres on the one hand and burlesque on the other to create the preconditions for the short story.[6] The North African's *Logic* offers a prescription of method sorely neglected today. His talisman approach is "to present broad things in a narrow way, narrow things in a broad way, ordinary things in a becoming way, new things in a familiar way, familiar things in a new way."[7] I hope to have shown by the end of this chapter that Blake's appreciation of Apuleius (and indeed Ovid) was neither entirely unconnected with significant series of paroles, nor with the fourfold root of the poet's Prophetic Books.

The metamorphosis theme does not play a great part in *The Fabillis* of Henryson. It was rather that, as Edwin Muir remarked, he "lived near the end of a great age of settlement, religious, intellectual and social" and "exists in that long calm of storytelling which ended with the Renaissance, when the agreement about the great story was broken."[8] Shakespeare is the great disruptor who bestrides that chasm and in what André Chénier called "ces convulsions barbares,"[9] metaphorically fused the old universalism and the encroaching individualism. The giant shadow of Shakespeare has seriously obscured the dimensions of a tradition that runs, in barest outline, from Aesop and Apuleius via the *Roman de Renart* and Henrryson through La Fontaine into the twentieth century with Masefield's grossly underestimated anti-War fable, *Reynard the Fox* and Dylan Thomas's ritual and fabulous tales, as far as Kafka and Nabokov. Apuleius had already pessimistically concluded in *On the God of Socrates*—"we men are the principal animated things, though most of us, through the neglect of training, are so depraved...through having nearly quite abandoned the mildness of our nature, that it may seem there is not an animal on earth more vile than man."[10] The moralitas of a Henryson fable can more comfortably be accomodated to a confident and medieval sense of human proportion, one in which the allegorical pattern of animal behaviour does not have the dangerous penetration of daily life present in Kafka's *The Metamorphosis*.

The problem is that animals follow a single line of development once on the road of growth. (The clouded tiger salamander is a complex exception, along with frogs). Insects, however, can metamorphose, destroying their former selves and rebuilding new identities. In Apuleius, there is a multiple structure in which the ethical lesson is entirely organic, partly at least because the morality has to be genuinely fought for within a fable where the type of transformations of identity resemble those of an insect, rather than animal. The first translator of *The Golden Ass*, William Aldington, warned that it may seem "a mere jest and fable," but that it is "a figure of man's life . . . a pattern to regenerate [men's] minds from brutal and beastly custom."[11] This is declared in the spirit that Keats characterized Shakespeare as leading "a life like the scriptures figurative" which in its allegory and mystery "people can no more make out than they can the Hebrew Bible." Likewise the Apuleian paradigm creates itself from precisely that extreme eclecticism and frequent reversal of style that Ben Edwin Perry had found deficient.[12] It is the product of Apuleius's instinctive imagination, what Perry reduced to "the fancy of the moment."[13] This impulsiveness leads to a certain unevenness of style and a tendency to stray or even contradict himself; but it is this spontaneous method that allows the metamorphic shape to emerge.

Apuleius stands at the crossroads of pagan and Christian civilization, as Pater's *Marius the Epicurean* dramatizes. Merkelbach insists that *The Golden Ass* belongs to those works that are "alles Mysterientexte."[14] But while taking on board this intent, Edgar Wind's description of the mixed nature of the ancient *grottesche* is also central:

> "Addressing the devout in a foolish spirit, these calculated freaks represented to perfection what Pico della Mirandola had defined as the Orphic disguise: the art of interweaving the divine secrets with the fabric of fables, so that anyone reading those hymns 'would think they contained nothing but the sheerest tales and trifles,' *nihil subesse credat fabellas nugasque meracissimas.*"[15]

This sly strategy is of a piece with the shifting identities of his character. For there is surely an association carried over from the Lucian *Onos* between *onos* and *onoma*, between donkey and name. To adapt Carlos Fuentes, Hermes "circulates names as if they were money and robs them of their permanence, which is the same as their essence," and this has spread to the names in *The Golden Ass* where "the number of nameless roles . . . far exceeds the named."[16] So Lucius can metamorphose into Apuleius himself in Chapter XI without disturbing the underlying unity of the book. Moravia has pointed to the source of the framentation of character as the pressure of an early mass society,[17] and the namelessness of so many expresses the already modern precariousness of individual identity in its development. It is as an anonymous and abused beast of burden that Lucius learns the sufferings of the exploitative world until his author rescues him to share the privileged, and expensive, initiations of Isis which are to re-establish his humanity on firmer foundations.

In the tale at the heart of the transformatory adventure, one critic has seen the very name 'Psyche' as the mainspring turning "the household tale into a beautiful narrative."[18] At first sight this may seem an extravagant claim, but taken together with the overall structure of the book, ψυχή proves indeed to be a key word and concept. The relative proximity to its folkloric origins allows the ordinary and extraordinary to merge in Apuleius's idiosyncratic but radical Platonism. The Scottish Lowland tale *Three Feathers* has, like its Hanoverian equivalent, many similarities to the Cupid and Psyche interlude in *The Golden Ass*. A girl is married to a husband she has never seen, and on lighting a candle she reveals him to be especially handsome. "But scarcely had she seen him when he began to change into a bird."[19] As Boberg tells us, the transformation of the man to an animal or bird is a Northern European and later development in this tale, when the divinity of the male has been lost.[20] In order to turn her husband back to a human, she must serve him for seven years and a day, which she does as a laundry maid aided by the three feathers her husband has given her. The story also includes a common theft element; so where Psyche tries to steal the beauty Proserpina has given Venus, the wife defrauds the serving men she works with of their savings. In the theft motif, there may be a sense of some illicit appropriation taking place in the recording of the tale, most particularly in the created Apuleian fable where the narrator is part of a band of robbers.

Carl Schlam has described many of the symbolic significances of the book in his *Cupid and Psyche: Apuleius and the Monuments*. But by drawing together some linguistic elements in Apuleius and relating them to the shape of *The Golden Ass* as a whole, I shall argue that an oblique approach best reveals the devious synthesizing quality of his imagination. The method is parallel to that sketched by the contemporary Spanish writer, Julián Ríos. In *Larva: Babel de una Noche de San Juan*, Ríos envisages the Spanish language as "a larval mask beneath which other languages lurk," while his book is "a *motamorphose* (word-metamorphosis)."[21] There is a cluster of words in Apuleius's *On the God of Socrates*:

> "Now of these lemures, the one who, undertaking the guardianship of his posterity, dwells in a house with propitious and tranquil influence, is called the 'familiar' Lar. But those who, having no fixed abode of their own, are punished with vague wandering, as with a kind of exile, on account of the evil deeds of their life, are usually called 'Larvae.'"[22]

Bearing in mind Psyche's escape from her family and sisters in marrying Cupid, and taking 'psyche' as soul including its manifestation as a butterfly on the monuments,[23] a fourfold archetypal pattern can be discerned. Norman O. Brown has caught a related group, from which I drew and then extended in Chapter 3:

> "*Larva* means mask; or ghost. *Larvatus*, masked, a personality—*larvatus prodeo* (Descartes); it also means mad, a case of demonical possession. *Larva* is also 'the immature form of animals characterized by metamorphosis'; in the

grub state; before their transformation into a pupa, or pupil; i.e., before their initiation."[24]

So there is a quadruple philological movement from ovum to larva, thence to pupa and finally psyche or butterfly, which is related to the spiral of education and maturation in the human cycle. Brown tends to see this as both an educational and also a characteristically American ("initiation") rite.

Apuleius's tale of transformation is also one of conversion in four stages. Judith K. Krabbe has isolated a sequence of Books I-III which follow Lucius's transmutation, Books IV–VI telling of Cupid and Psyche, Books VII–X revealing the darker perhaps chrysalitic stage in which indeed gold, *chrusos*, becomes the main arbiter of Lucius's destiny, followed by the regenerative Eleventh Book.[25] Although Apuleius has often been criticized for being more of a populariser than a philosopher, the fact is his instantaneous understanding of the process of individual growth and development surpasses that of his master, Plato, whose *Phaedo* he translated. Plato conceived a soul that was feathered,[26] but Apuleius envisaged the soul winged as a butterfly, whose form in evolution *The Golden Ass* as a whole mimics. (The centrality of mimicry among lepidoptera in the development of Wallace's and Darwin's theory of evolution, as well as its contemporary place in the writings of the lepidopterist Vladimir Nabokov, will be the subject of a later chapter.) We thrust back to the sixth century BCE and earlier when the soul was designated a butterfly even before it was personified.[27]

Within the context of 'Cupid and Psyche,' it is upon the completion of her fourth task that Psyche earns the right to rejoin her husband. P.G. Walsh admits that "there is something in the thesis of Reitzenstein . . . who compares eastern versions to propose that the visit to the underworld is not merely a fourth task but the organic close to the story, by which Psyche gains heaven."[28] Four tasks appear in African folk-tales as compared to the more usual three and it is significant that salvation in the case of Apuleius's middle demons, as he terms them, is so inextricably associated with the quaternal. Indeed this is, *pari passu*, repeated in Raphael's scenes in the Farnesina cycle— the very cycle so pedantically recommended to Dorothea by Casaubon— where the six scenes of divine displeasure are succeeded by the four scenes of Psyche's reinstatement and ascension. This swathe of numerical symmetry and complex linguistic inference lies behind Apuleius's work and animates the totality. It ultimately opens up a concundrum in the sphere of Evolutionary Theory that literature and the arts, rather than science, may help to resolve.

Joseph Conrad's favourite reading was A.R. Wallace's *The Malay Archipelago*, and Richard Curle remarked on the special place this contemporary theorist—"more Darwinian than Darwin himself"[29]—had in Conrad's affection: "He had an intense admiration for those pioneer explorers—"profoundly inspired men" as he called them—who have left us a record of their work; and of Wallace, above all, he never ceased to speak in terms of enthusiasm."[30] Although Wallace is not the model for Stein in *Lord Jim* as Florence Clemens would have it, nonetheless the sense of "the accuracy, the harmony," and primarily "the beauty" of Stein's collection of butterflies may well have arisen from Wallace's exclamations at the splendour of *Ornithoptera croesus* when he felt fainter than when threatened with death, and from the rapture he felt at the "fresh and living beauty" of another of the large bird-winged butterflies.[31] It is clear, as Norman Sherry argues, that many details making up Stein came from Dr. Bernstein, Charles Allen (once a youthful assistant to Wallace), and Captain Lingard. But in regard to the light thrown on Jim's life by Stein's ruminations on lepidoptera's interweaving with human life, Wallace is the central candidate for the thrust of this outlook. Moreover, the episode in which Jim is imprisoned in Patusan parallels Wallace's confinement on entering the village of Coupang during his visit to Lombock and Bali, while the whole line of development of the novel follows the actions and experiences of Darwin's co-founder of natural selection.[32] So Conrad has cast aspects of Wallace's entomological bias as having a complex philosophical significance, a significance Darwin's work lacked except in the negative sense as a shattering of religion.

Darwin notoriously lost his responsiveness to music and most literature as his life progressed. But Wallace always insisted that mathematical, musical and artistic faculties intimated "the existence in man of something which has not derived from his animal progenitors . . . a spiritual essence or nature, capable of progressive development under favourable conditions."[33] He seriously harmed his alternative cause through his excursions into spiritualism and such impossibilisms as the "projected development of spiritual beings capable of infinite life."[34] And in his reading of Swedenborg, he enthusiastically marked a passage declaring that "it must be known that all spirits and angels without exception were once men, for the human race is the seminary of heaven . . ."[35] Nevertheless his concentration on insects, together with his insistence that the nature of the human brain lay in "the *evasion* of specialisation" suggest a more profound gulf than either Wallace or Darwin found it tactically opportune, in the light of the newness of the central theory and the opposition aroused, to admit between their views.[36] As I have said in another chapter, an approach that gives full weight to the part played by literature and the arts in the nature of humanity yields a view of the species as poised, or perhaps merely Bellow-like dangling, between animal and insect. While great ape ancestors are indisputable, the cultural evolution of the individual is better understood in terms of the life cycle of the metamorphic insects, a process that both masks its future self—the root *larva*—and then destroys its former self in order to proceed to the next stage of growth. Wallace in fact opposed "the inclusion of man's psychical nature as a product of evolution,"[37] but this left him in his later quagmire. It can be said that the imagination persistently strives to assist the human species to understand its evolutionary position by seeking out its associative root, *imago*. In this the writings of Apuleius are a real advance not just on Plato, but also on Ovid's *Metamorphoses*. For although the opening of Apuleius's *Metamorphoses* literally echoes Ovid,[38] his vision of transformatory conversion surpasses his predecessor's less organic conceptions. Here Pythagorean mathematics has been reanimated through a magical, or at least mysterious, refusion with nature's processes.

NOTES

1. Reitzenstein, 38 and 68ff.
2. Márquez, 112.
3. Rushdie, 301.
4. Eliot, 229.
5. Damon, 26.
6. Coburn (2), 4:5463. The Apuleian origin of the short story is proposed by Ben Edwin Perry, 248 and 282.
7. Londey, 83.
8. Muir, 10.
9. Chénier, 647.
10. Apuleius (3), 354.
11. Apuleius (1), xxxviii.
12. Perry (2), 249.
13. Perry (2), 240.
14. Merkelbach, 89.
15. Wind, 237.
16. Fuentes, 182.
17. Moravia, 172.
18. Purser, liv. I should add that Purser opposed the allegorical interpretation of Psyche as soul, just as Nabokov abhorred the symbolic transposition of butterflies.
19. Briggs, 1:511.
20. Boberg, 1:216.
21. Levine, 182; Gautier, 185.
22. Apuleius (3), 364.
23. Schlam, 32.
24. Brown, 96–97.
25. Krabbe, 70

26. Plato, 487ff.

27. Immisch, 193. Without obstructing the continuity of my argument in the body of the text, a section of the Immisch is a crucial addition. "Als Gesichert wird vorausgesetzt, daB auch Seelenschmetterling eine Abart des sogenannten Seelenvogts ist, nicht als das einzige Insekt. Es steht also am Anfang etwas, was weit entfernt ist von dem ammutigen Elfentum der spätren Faltermädchen und was sonst von zierlicher Symbolik in Frage kommt. Seelenvögel sind ein unheimliches Gelichter."

28. Walsh, 212–13.

29. Wallace (5), 2: 22.

30. quoted Sherry, 142.

31. The first two quotations are from Conrad, 208. The final one from Marchant, 1:22.

32. Further parallels are traced in Houston, 30–45.

33. Wallace (5), 2:474.

34. Wallace (5), 2:477.

35. Swedenborg (1), 12.

36. Eisley (1), 306.

37. Clodd, 134.

38. Krabbe, 43–44. Brooks Otis has controversially distinguished a quaternal plan in Ovid's *Metamorphoses*. In *Ovid as an Epic Poet* (Cambridge: Cambridge University Press, 1966), 83.

Chapter Six

The Greek Supplementals in Their Connections to *ΨΥΧΗ* or How Breath, Soul and Butterfly are Linguistically Interconnected

In Apuleius's tale of 'Cupid and Psyche' within his *Metamorphoses*, the Soul and Love combat their way towards unity, although the allegory is not insisted on throughout. This is a tale told in Latin, but the philosophical and linguistic roots of the story lie in the Greek—philosophically in Plato's *Phaedrus* and linguistically in the word ψυχή.[1] Psyche is a key word in Greek since it evolves from Homer to signify not only 'breath of life,' but also mind and soul, and is ultimately imaged as a butterfly. Nietzsche's friend Erwin Rohde gave a magisterial if idiosyncratic account of this word in Greek culture some 110 years ago in his *Psyche: The Cult of Souls and Belief in Immortality among the Greeks*. The composition of the Greek word psyche, ψυχή, is intriguing. The letters ψ (ps) and χ (kh) have an interesting history. They are two of the three so-called supplementals added by the Ancient Greeks to the Canaanite Semitic script adapted from a syllabic language to create the first alphabet which we in the West have, of course, inherited. Φ, χ and ψ do not appear in Phoenician, and were added by the Greeks.[2] The supplementals belonged to the earliest alphabet, but at that stage ψ represented the sound 'kh'. The first set of supplementals arrived among the Euboians. Later, a reformer, an Ionian, clarified χ as kh and ψ as ps, discarding φσ as 'ps' and χσ as 'ks.' While Athens retained these older forms, Corinth took up the new symbolic significances. Later the modified letters became part of the Koine script. So these were aspirates which came to form the core of the word ψυχή, itself signifying, inter alia, 'breath.'[3] Eric Havelock sees invention by the Greeks of the alphabet as dissolving "the syllable into its acoustic components—we might almost say its biological components in so far as these are actually effects produced by movements of different parts of the human body."[4] Bickel argued that the Homeric ψυχή did indeed represent the very process of respiration.

45

Barry Powell argues, following H.T. Wade-Gery, that "the signs ψ and X introduce confusion, not clarity,"[5] and yet here he forgets his main thesis that the literature of Greece produces the language through the adapter who wrote down the hexameters of Homer. (Havelock's term 'inscriber' for the adapter is a more appropriate term.)[6] Again, Powell insists it is not at all likely the three supplementals were invented as a result of evolutionary need.[7] However Andrey Bely offers a different perspective on this. In his essay "The Magic of Words," Bely points to the role of literature in creative cognition:

> "When I assert that creation precedes cognition, I am asserting the primacy of creation, not only because creation is epistemologically superior, but also because it is prior in actual genetic sequence."[8]

So in the two signs under discussion, subjective intuitions of the Greeks inspired their invention—intuitions that finally reached the arena of literature in the Latin tale of Cupid and Psyche, which had been intimated in the *Rig Veda*.[9] When Powell writes that ψ and X will have belonged to the original system because "we would expect new signs added to a preexisting signary to clarify ambiguity," and that "the signs ψ and X introduce confusion, not clarity," he is making assumptions about the rationality of alphabetic introductions that misunderstand the "genetic sequence" to repeat Bely's fertile phrase.[10]

Letters of the alphabet were defined by the Greeks as 'voices', that is to say that the *sound* was primary. Indeed Powell makes the pertinent point that Archaic Greek "in its obsession with phonetic accuracy . . . was a great anomaly in the history of writing."[11] As W.B. Stanford put it: "Written words were more like memory-aids to remind readers of certain sounds . . . The Greek grammarians emphasized 'orthoëpy,' the correct rendition of texts "according to intonation, timbre-quality, and quantity, as the prerequisite of literary appreciation."[12] So these supplementary signs were necessary to the Greeks as elements to be assembled for the concise expression of their concept of the soul, which became a form of mimesis through the breath sound of ψυχή. This is in line with the remarkable scientific accuracy of their alphabet:

> "The inventors of the atomic theory of matter were the first possessors of a system of writing where graphemes represent the 'atoms of spoken language,' an analogy explicit in the Greeks' use of the word *stoicheion*, 'something in a row,' both for an alphabetic sign and for an atomic element. Greeks imagined that the structure of their writing parallelled the structure of the phenomenal world, according to the unobvious theory that matter consists of a limited number of discrete particles invisible but real, which act in combination to produce visible effects."[13]

The Indo-Europeans generally depended on syllable counting; the Greeks on length, short and long. Boisacq distinguishes a provisional Indo-European etymology in *bhes- (souffler) leading onto ψυχή and Bickel interprets the Greek word as "breath-soul".[14] On the other hand Claus insists throughout his book *Toward the Soul: an inquiry into the meaning of ψυχή before Plato* that it not so much signifies 'breath' as 'life-force' together with an intimation of coldness. However there is an unusual confirmation of the *origins* of the nuances of the word. Recent medical research has been carried out into the relation between reading Homer aloud and cardiovascular health:

> "The effects of different breathing frequencies and patterns found in poetry readings on cardiovascular regulation have been investigated extensively in recent years. Poetry recitation has been known to cause a frequency adjustment of breathing oscillations with endogenous blood pressure fluctuations (Mayer waves) and even cerebral blood flow oscillations during the saying of the Catholic Rosary and the 'OM' mantra.
>
> The effect is attributed to the breathing frequency of approximately six breaths per minute induced by the metric of both religious verses . . . so that heart rate and respiration may intermittently synchronize."[15]

The product of these originally oral poems of Homer consists both of an impact of poetic thought and at the same time a guarantor of actual health, utilitarian and aesthetic in result as well as an affirmation of the ancestral traditions of the community of listeners and speakers. The long and short syllables of the hexameters of the *Iliad* and *Odyssey* are particularly beneficial. During recitation of hexameter verse the low frequency oscillations of the breathing pattern were synchronized to a large extent with the heart rate oscillations. The six metres or rhythmic units per line slows the normal breathing rate from about 15 breaths a minute to just 6. This in turn synchronizes with the regular fluctuations of blood pressure which normally go in ten-second cycles. Powell asserts that the dactylic hexameter "is controlled by a structure 'deep' in the user's psyche."[16] Perhaps in the light of the recent scientific research, 'diaphragm' could be substituted for 'psyche' in this instance. Chorus and audience would recite more than 10,000 lines without pausing at performances in Ancient Greece.

Bruno Snell has pointed out that Homer's uses of *psyche* are not individuated in the modern way, that the *psyche* in no way belongs to the person. Achilles' reactions "are not explicitly presented in their volitional or intellectual form as character, i.e. as individual intellect and individual soul. Mental and spiritual acts are due to the impact of external factors, and man is the open target of external forces which impinge on him, and penetrate his very core."[17] This receptivity to the outside impressions again gives an added objectivity to the ψυχή, which will later in history take on an independent existence as an insect whether a butterfly as in Estonia and Ireland, or bee, wasp or dragonfly in Japan. The soul is in this sense purely objective; it is the breath of life and is quite distinct in life from the person. This objectivity arises from the very apparent weakness in psychological portrayal of Homeric heroes, their failure to distinguish between what was inside and outside themselves as moderns do. At death the *psyche* departs the individual for Hades, or is loosed. It is this sheer objectivity, the non-personal soul, that makes the Homeric epic so rivetting and noble, arousing a sense of truth to life. The incorporation of the *psyche* into personal psychology does not take place until the end of the fifth century when a person is conceived as relating to his or her psyche in the course of a life of independently selected activity. In Homer the *psyche* is a phenomenon in a state of transition. It is likely there was a word for the free soul which in Homer is gradually replaced by the soul as 'breath of life' while at the same time beginning to lose its purely physical limitation. Despite the remoteness of concepts—*because* of the remoteness—there is an awful power to the *Iliad* and *Odyssey*. As Snell puts it:

> "The *psyche* [in Homer] leaves through the mouth, it is breathed forth; or again it leaves through a wound, and then flies off to Hades. There it leads a ghostlike existence as the spectre (*eidolon*) of the deceased. The word *psyche* is akin to ψύχειν, 'to breathe', and denotes the breath of life which of course departs through the mouth; the escape from a wound evidently represents a secondary development."[18]

When Andromache swoons, Homer says that she "breathed forth" (ἐκάλυψεν) her ψυχήν (*Iliad*, 22, 467), a word "most likely connected with smoke" as Bremmer remarks.[19] All other Greek words connoting aspects of the individual soul are also connected to 'breath'—*thymos*, *noos* and *menos*. The paradox is that it is only at the instant of death that the Homeric psyche becomes evident. During the life of the hero it is implicit, for his identity is determined by his organs. Yet at death psyche has the muscular strength of an organ, an entirely objective entity in no way dependent upon subjective states. This projects the human into a special relation with nature, which is defined as Fate, while the word ψυχή also later reaches across to be imaged as a butterfly. By the time of Hesiod and the lyric poets, it has drawn closer to the psyche element in our own concept of psychological but the concept of the grandeur of the human has begun to shrink.

R.B. Onians though delves into the pre-Homeric origins of ψυχή by way of Plato's *Timaeus*. This will suggest the physiological roots of the sounds of the word, akin to the snorting that John Burnet rather daringly proposed in his pioneering address to the British Academy some 90 years ago.ago.[20] Onians argues that ψυχή is located "in the marrow, the divine part in the marrow of the head called ἐγκέφαλος . . . The ψυχή is itself 'seed' (σπέρμα), or rather is in the 'seed', and this 'seed' is enclosed in the skull and spine . . . It breathes through the genital organ. This appears to be original popular belief."[21] For Aristotle the seed was itself breath or had breath (πνεῦμα), and procreation itself was such a breathing or blowing.[22] And so the early psyche is both more and less than the 'breath-soul'. It is not lung-related, but in a sense is more profoundly related to sexual generation. It suggests an erotic root for our word, and this is confirmed by the Dionysiac image on a black amphora from the sixth century BCE which shows a butterfly beneath four drops of semen from a dancing satyr.[23] Then again Hermes was represented as a squared pillar head with genital organs, while an Etruscan scarab shows him with a butterfly on his shoulder.[24] Here the psyche is fertilized with generative powers, so representing the new epoch of wealth. And so the circle of meanings of ΨΥΧΗ close, offering life-force together with breath-soul.

NOTES

1. Harrison, 256–58.
2. Powell (1), 55.
3. Claus rejects the interpretation of the Homeric psyche as 'breath of life' and plumps for the Nietzschean and Shavian 'life force.' (Claus, 7, 92–97). He sees Homer as purging the word of its possible (as he sees it) original meaning of 'breath', and replacing it with "the 'life-force' category of words (97)."
4. Havelock (1), 69.
5. Powell (1), 57.
6. Havelock , 33, fn.18
7. Powell (1), 49.
8. Bely, 93.
9. Lang, 66.
10. Powell (1), 57.
11. Powell, "Homer and Writing" in Morris, 4.
12. Stanford, 3, 7.
13. Powell (2), 23–24.
14. Bickel, 49.
15. Cysarz et al.
16. Powell (1), 224.
17. Snell, ix, 20.
18. Snell, 9.
19. Bremmer, 22.

20. Burnet, 245.
21. Onians, 119.
22. Onians, 120.
23. Schlam, 7.
24. Schlam., 7.

Chapter Seven

Coleridge's Tetraktys: Summit of His Metaphysics, Death of His Poetry?

All the signs are that in today's intellectual milieus, poetry is more highly estimated than philosophy. This is partly because of the attempts to take poetry to a wider audience than half a century ago, but also the terrific status that immediate experience has in late twentieth and early twenty-first century culture. However, Coleridge lived on the borderline of poetry and philosophy, taking Occam's razor to his poetry faculty because he realized that to pursue the questions that motivated him throughout his career, he had to plunge more deeply into philosophy. He had to abstract from his own and other poetry in order to attain that world-view that grasped the cosmos as the emanation of a single principle, and so as a unity. The key to this was in understanding the growth of the mind when it falls under the influence of the imagination, and in comprehending how this relates to the outer world. So Coleridge hinted that the processes of the growth of mind were coterminous with the forces of growth in nature. Without the poetic imagination, it was impossible to evolve an exact metaphysics. But in formulating the metaphysics, a heavy personal price had to be paid. As he lamented in a letter to Southey, his ability to create poetry had declined as his "long and exceedingly severe metaphysical investigations" deepened.[1]

Coleridge's *Hints to the Formation of a More Comprehensive Theory of Life* was written rapidly during November and December 1816. It thus follows on the heels of his *Biographia Literaria* of 1815 (though its Preface dates from 1800), as he turned away from the German philosophers. Indeed one critic has characterized this apostasy as "nothing less than a philosophical rout, a mad dash from an untenable position."[2] In precipitating himself away from Schelling's *Natur-Philosophie*, he began to sketch a pantheism in which nature was "no mere finished and dead product (*natura naturata*), but

51

... a living and creative principle (*Natura naturans*)."[3] In making this transition, he made more specific in his Lectures some of his evolutionary ideas on insects, whose nervous system and whose function as antennae of the health of an ecology, philosophically fascinated him:

> "Nature takes a higher step and passes into the fishes [from insects], and there the nervous [system], the object by which reflection or memory is rendered, [probably] begins. But in doing this she has lost something; all the [instincts] of life, all that delight in the instinct is gone (*Collected Works* (*CW*), 8, i, 384–85)."

More exact than D.H. Lawrence's sense of instinct in being part of a working philosophical system, nonetheless this sounds distinctly Lawrentian. But it was in his *Hints* that he gave his fullest account of the entomological universe:

> "The insect world, taken at large, appears as an intenser life, that has struggled itself loose and become emancipated from vegetation, *Florae liberti, et libertini* [Fora's freedmen and their children]. If for the sake of a moment's relaxation we might indulge a [Erasmus] Darwinian flight . . . we might imagine the life of insects an apotheosis of the petals, stamina, and nectaries, round which they flutter, or of the stems and pedicles, to which they adhere. . . All plants have insects, most commonly each genus of vegetables its appropriate genera of insects; and so reciprocally interdependent and necessary to each other are they, that we can almost as little think of vegetation without insects, as of insects without vegetation. Though probably the mere likeness of *shape*, in the *papilio*, and the papilionaceous plants, suggested . . . this fancy of a flying blossom; when we reflect how many plants depend upon insects for their fructification (*CW*, 11: i, 541–42)."

He concludes that the insect world is characterized by irritability—presumably because of the presence of a nervous system—while the vegetable world is defined by reproduction. Drawing on the work of a follower of Schelling, Heinrich Steffens, Coleridge suggests that the metamorphoses of insects "are but an individuated and intenser form of a similar transformation of the plant from the seed-leaflets, or cotyledons, through the stalk, the leaves, and the calyx, into the perfect flower (*CW*, 11: i, 545)." Later he quotes a passage from Pierre Huber's *Natural History of Ants*, describing the subtle behaviour of the ants in the construction of the walls of their chamber. Coleridge attributes quite a high level of development to the insects when he asserts that "the faculty manifested in the acts here narrated does not differ in *kind* from Understanding, and that it *does* so differ from Reason (*CW*, 9, 222)."

So insects have the quality of intelligence and reflection, according to Coleridge, yet lack the ability to systematize that Reason allows in its path to truth. What he intends here is that ants and bees are more capable "of adapting means to proximate ends than the elephant (*CW*, 14: ii, 80)." And the editor of this 1830 volume, *Table Talk*, recalls that the poet "was accustomed to consider the ant as the most intellectual, and the dog as the most affectionate, of the irrational creatures, so far as our present acquaintance with the facts of natural history enables us to judge (*CW*, 14: ii, 80 n.)." If we follow through the logic of Coleridge's train of thought, then the august Immanuel Kant is reduced to the level of insect since his philosophy only attained the category of Understanding. Kant's rejection of the ability of mind to define the noumenon, to know the thing-in-itself, objective reality. This meant that Kant could not achieve the human capacity "to know ultimate truth" because "understanding operates only on the experience of the senses" while "the moral necessities of reason are supersensuous (*CW*. 14, i, 39 n.44)." It is to insects that he turns for traits resembling not only human understanding and affections, but also moral behaviour:

> "Insects and animals are with few exceptions applied in malum usum, to the evil passions and appetites—& where otherwise, as in the instance of the Ant and Bee, yet not to the Affections, but to the prudential obligations, to the exercises of the Understanding as the potential Adaptative Faculty, the intelligent *Instincts* in the Animals—this being the Flower & culmination of the 2nd Form of Life, the Insectivity, most inadequately called Irritability—(*CW*, 12: ii, 689)."

Coleridge's introjection of the insects into the history of philosophy is inspired, and if he had a little more of Apuleius's sense of humour then he could have taken that history into the realms of the burlesque. The sense of many of the great philosophers, always excepting Pythagoras and Plato, straining at a gnat is palpable.

So having outstripped Schelling and Kant, then by 1818, he reflected in his Notebooks that "the more I reflect, the more important do both the Pythagorean or arithmetical, and the Platonic or harmonical Schemes of Nature present themselves to my mind."[4] It is back to the pre-Socratic Pythagoras that the poet especially gravitates. For Pythagoras—the first who can justly be called a philosopher, according to Coleridge—"found a relationship between the Soul of Man and the laws of Nature—he perceived that what in Man is the idea, is also in external Nature the law . . . (it was in figures & numbers he found a certainty beyond contradiction and in nature the same unity was explained by his mind) (*CW*, 8, i, 101)." Later, probably thinking of his beloved Aeolian Harp, he gives this a more personal and poetic slant—

> "We have, alas! Nothing more than the faint Echoes of the Pythagorean Doctrines, vibrated to us from the distant Antiquity. Yet the more carefully I [erased]

we have collected the fragments, . . . the less possible does it become to regard the minimal Symbols of Pythagoras as mere fancies or accidental coincidence, the more do we feel disposed to *worship the echo. (CW,* 11, i, 437–38)."

The Tetractys becomes the symbol of Nature and here once again we have, not just as a random interest but a mature conclusion, the Cosmic Constant reappearing. Indeed apparently in his own formulation, Coleridge ventures so far as to invent the conception of the "Four-wheeled life, the four-wheeled *vehicle* of life (*CW,* 11: i, 438)." He dismisses Schelling's *Philosophie und Religion* declaring the German thinker "*toils* in and after, like the Moon in the Scud and Cloudage of a breezy November Night" what "is far more intelligibly and adequately presented in my Scheme or Tetraxy (*CW,* 4: 400–01)."[5]

The whole system elaborated by Coleridge echoes Pietro Bongo's 1585 assertion that "from the quaternions arise the roots of the whole world."[6] This is the Cosmic Constant resurfacing after a period of submergence. Alastair Fowler gives an account of a key chapter of Bongo's *Mysticae numerorum significationis liber:*

"Having cited Plato's view that the soul is a harmonic number composed from the *tetractys,* Bongo surveys several theories of the four-fold nature of the soul; such as the theory that there are four souls *veluti fundamenta* [as though the bases]: namely *mens* [mind], *scientia* [knowledge], *opinio* [opinion], and *sensus* [perception]. The Pythagoreans, he notes, postulate a different correspondence between the soul and the quaternion; for they say that the soul is composed of four spiritual elements, and that it exerts a four-fold spiritual force: concupiscible, irascible, rational and voluntary."[7]

But it is when Coleridge comes to define the God, that he felt the Pantheists had compromised, that he asserts "the adorable Tetractys, or Tetrad is the formula of God, which again is reducible into, and the same reality with the Trinity(*CW,* 14, i, 288–89)." The fourth element, or Prothesis, assures the unity of Father, Son and Holy Ghost. So, the Trinity "must be contemplated in the Light of the Tetractys (*CW,* 12, ii, 745)," and Father, Son and Spirit "must be contemplated as four (*CW,* 12: ii, 745, n.92)." There is "a *Tetrad* in *divine matters,* a *Pentad* in all others (*CW,* 12, v, 631n)." At other times, the pentad is represented as God's hand in the world, and indeed Kiyoshi Tsuchiya has distinguished a fivefold pattern of regeneration in "The Ancient Mariner." But the primary and fundamental value for Coleridge, that which leaves him at home in the universe, is the tetrad, the tetractys. Coleridge has in a sense sacrificed his poetry to make his way through the labyrinth to this conclusion, which renewed the Cosmic Constant, spanning back to the medieval and ancients.

NOTES

1. Coleridge (2), 1:388.
2. McFarland, 156.
3. Muirhead, 118.
4. Coburn (2), 3:4436.
5. "Tetraxy" is Coleridge's coinage for Tetractys, implying his familiar relationship to it. It does not appear in the *OED*.
6. Alastair Fowler (1), 277.
7. Alastair Fowler (1), 277.

Chapter Eight

William Blake,
Rhapsodist of the Fourfold

As the title of this chapter suggests, William Blake singing a song of himself is, like Whitman, singing of the world as a whole.[1] He excavates the recurrent pattern that manifests itself in works of the imagination, but which because of the frequency of its emanation clearly arises from an embedding within the human organism. If there is to be a replicator parallel with genetic encodement, then it must contain this. If the problem raised by Maynard Smith is to be overcome as to the mechanics of transmission, it will surely have to be addressed in this arena. The most likely solution is that an imprinted system is at work, but one functioning with the assistance of the whole of the organism. Cavalli-Sforza remarks that "in evolutionary biology, 'transmission' generally connotes Mendelian genetic transmission, and studies of evolution under this mode of transmission constitute a large part of population genetics. Much less has been written about the evolution of traits whose transmission is not genetic."[2] Blake's opposition to Locke is significant here. For Locke, the human mind is a *tabula rasa*. There are no innate ideas, but only sense impressions. Blake on the other hand assumes that humanity embodies the whole cause of matters, and that the systole and diastole of life expresses that. In arguing for the imprinting of what I am calling the universal grammar of the great imaginative constructions from literature to religion, a proof has first to be imagined. As Blake wrote: "What is now proved was once only imagined (*MHH*, III, 33)."

Northrop Frye demanded that "in Blake all recurrent numbers and diagrams must be explained in terms of their context and their relation to the poems, not as indicating in Blake any affinity with mathematical mysticism."[3] The sophistry in this academic pronouncement becomes especially clear if Blake's vision is related to the universal grammar of imaginative invention. The building of Jerusalem, "The great city of Golgonooza," is "fourfold" and "perfect in its building, ornaments & perfection. (*J*, 12.46, 53)." As one critic puts it: "for Blake, the ultimate city is Golgonooza, a name combining the place of Christ's sacrifice with the primal ooze of existence."[4] In his *Milton*, he is even more explicit that it is constructed from the building blocks of creativity, and has to be continuously invented in the midst of nature:

"Golgonooza, the spiritual fourfold London eternal, In immense labours & sorrows, ever building, ever falling, Through Albion's four forests which overspread all the earth—(M, 6.1–3)."

And he goes on to relate this to the unearthing of the human in humanity, this actually seen as connected to the insect:

"O how can I, with my gross tongue that cleaveth to the dust,
Tell of the fourfold Man, in starry numbers fitly ordered? . . .
Now Albion's sleeping humanity began to turn upon his couch,
Feeling the electric flame of Milton's awful precipitate descent.
Seest thou the little winged fly, smaller than a grain of sand?
It has a heart like thee, a brain open to Heaven & Hell,
Withinside wondrous and expansive. Its gates are not closed;
I hope thine are not. Hence it clothes itself in rich array;
Hence thou art clothed with human beauty, O thou mortal man."
(*M*, 20.15–16, 25–31).

Jerusalem—for which read the universal grammar of imaginative creation—stands just outside the human heart, in the midst of destructive chaos, constantly reconstructing: "And every part of the city is fourfold, & every inhabitant fourfold (*J*, 13.20)."

This quaternity is both inside and outside humanity, both subject and object. Humanity has to draw upon it without rest:

"Here, on the banks of the Thames, Los builded Golgonooza,
Outside of the gates of the human heart, beneath Beulah
In the midst of the rocks of the altars of Albion. In fears
He builded it, in rage & in fury. It is the spiritual fourfold
London—continually building & continually decaying desolate."
(J, 53.15–19)

Blake's conception is closer to permanent intimations of the universal grammar of creativity than any other world poet. It does, however, have its roots in Milton's *Paradise Lost*. There, as Röstvig points out:

"To the Renaissance mind, however, the fourfold nature of the world and of the divine scheme of redemption was a *datum*, an absolute truth contained in the Bible as interpreted by all the great authorities. For this reason the fourfold structure of Lamentations was a prophecy of the 4 gospels to come. That Milton should have used this number for structural purposes, therefore, would have lifted no critical eyebrows in 1667. Indeed, on coming across the centrally placed episode of the ascent in the chariot, readers would have been reminded of Christ's position in the middle and of the way in which he imposes harmony through the number 4."[5]

Blake has embodied this in a series of gigantic evolutionary epics. The cities of Britain gather to assist the lost Albion:

"At length they rose
With one accord in love sublime, & as on cherub's wings
They Albion surround with kindest violence, to bear him back
Against his will through Los's gate to Eden. Fourfold, loud,
Their wings waving over the bottomless immense, to bear
Their awful charge back to his native home." (*J*, 38.82–39.5).

Jerusalem mourns the fact:

"Once a continual cloud of salvation rose
From all my myriads; once the fourfold world rejoiced among
The pillars of Jerusalem, between my winged cherubim,
But now I am closed out from them in the narrow passages
Of the valleys of destruction, into a dark land of pitch & bitumen,
From Albion's tomb afar and from the fourfold wonders of God
Shrunk to a narrow doleful form in the dark land of Cabul." (*J*, 79.57–63).

Stuart Curran has concluded that "the Neoplatonism that Kathleen Raine finds so prevalent in Blake's writing is more correctly Christian gnosticism."[6] Raine tries to tie Blake's Eternals into pagan tradition. As Peter Sorensen points out, his Eternals are built around the idea of a Tetrad, with forces equally distributed among the 4 'zoas' or beasts, and such beastly tetrads appear in the Bible quite frequently. He further remarks:

"In gnosticism, the four gods corresponding to the Eternals may be the 'four Light-givers, who stand in the presence of the Great Invisible Spirit',(*Nag Hammadi*) . . . The four Eternals are not the only godlike figures in Blake, nor do the 'four Light-givers' represent the only godlike figures in ancient gnosticism. Yet it is important to find confirmation in gnostic sources of the Blakean tetrad."[7]

The Gnostics were centred in Alexandria and sought to unite Greek philosophy with eastern science and religion on the basis of Pythagoreanism. For them, the intellect is based upon a few simple numbers, which recur in all philosophies. Vincent Foster Hopper explains that:

"As a result of the infusion of Neo-Pythagorean elements, 3 and 4 became the basic numerical symbols of the theology [of Gnosticism]. As in Pythagoreanism, all proceeded from the One, but the sensible world was produced from the triadic haromony of Being, Life, and Intellect in association with the 4 creative elements or the 4 elementary kingdoms. The sanctity of the 3 received new emphasis from the triadic groupings of Egyptian and Babylonian gods."[8]

Further there are remarkable parallels between Blake's philosophical system and Buddhism, something perhaps set off by his reading of the first English translation of the Bhagavad Gita. He himself envisaged a connection between his thought and the Eastern systems. There are 4 important events in the way of the Buddha—birth, enlightenment, the first sermon and death—and four of lesser import: the descent from heaven, multiplication of himself, the taming of the elephant and the incident of the bowl of honey offered by monkeys. The doctrine of the Four Noble Truths is an integral part of Buddhist scriptures, such as the *Digha Nikaya*, the *Majjhima Nikaya*, the *Sutta Pitaka* and the *Vinaya Pitaka*. As D.P. Singhal writes:

"Whilst in Western civilizations the interest has increasingly focused on single sciences, in the Indian world the ontological viewpoint has been generally preferred, and it would thus appear that 'in India, through all periods, the special sciences are rooted in and developed on the underlying unifying cosmic concepts and presuppositions of which the single scientific result is only a special case and phenomenon, a demonstration and a facet, as it were, of the universal cosmic law. The universal vision in India has never been lost.'"[9]

The Gita though claims a man can only become immortal when he liberates himself from his gunas or bodily expression, while Blake sees no separation of body and soul as the Four Zoas enact regeneration.

The disintegration of humanity is the disintegration of the fourfold, a splintering into fragmentation. In Blakean mythology, the loins are Tharmas, the heart Luvah and the brain Urizen. They are unified in the mind of Urthona, so that there is a fourfold organism. Without the coming together, there is distortion, in genetic terms a phenotypic malfunction. But Blake goes further than this in a formulation of creativity's skeleton. The unification of all 'elements' in the human is "consummated in mental fires" because the knowledge of this unity is the supreme act of the Imagination, at the point where it becomes "the Human Existence itself (*VFZ*, 7, 349–52)." Frye offers an explanation of two competing numbers:

"The numbers four and three in *Jerusalem*, as in *The Four Zoas*, are respectively the numbers of infinite extension and cyclic recurrence. *Jerusalem* has four parts, and the end of the third brings us back to where we started. Numbers built up of fours, such as sixteen and sixty-four, signify imaginative achievements, like the Bible and Golgonooza; numbers built up of threes, notably nine and twenty-seven, are associated with Antichrist. Similarly in the Book of Revelation, where the sacred or imaginative number is seven, the Antichrist is represented as a series of sixes. Twelve, the product of four and three, is the number of humanity as it is born into the fallen world, with its imaginative and natural tendencies fighting one another for its soul."[10]

In Blake the domain of the threefold is Beulah, "a soft moony universe, feminine, lovely," designed for creatures who could not sustain the mental fight required by Eden. For Blake it is only a place of rest, of threefold life, not of the highest fourfold life of Eden ("Thus wept they in Beulah over the four regions of Albion (*J*, 25: 14))." Through Los and Enitharmon though, the two numbers can be allied as again in Eden:

> "If we unite in one, another better world will be
> Opened with your heart & loins & wondrous brain,
> Threefold as it was in Eternity & this the fourth universe
> Will be renewed by the three and consummated in mental fires."
> (*VFZ*, viia, 349–52).

In fact Blake's vision of the world is that it is a reflection of the quadruple. There are the four faces of humanity, Urizen, Los, Luvah and Tharmas, and when these are in conflict with each other chaos ensues as each acts autonomously. Humanity itself is divided into the four: Humanity which synthesizes, the Spectre which separates individual from individual, the Emanation which is the visible and spatial aspect of the individual, and the Shadow, or remnant of desire after the decay of passion and fire. There are four visionary kingdoms: Eden, Beulah, Ulro and Generation. And history is a product of fourfold man: Africa, Asia, Europe and America. Africa is where the Fall and History begin; Asia where capitalism originates; Europe where the abstraction of Ulro reigns; and America where regeneration begins.[11]

Blake's spat with Newton's theories is provoked by his concept of a mechanical universe susceptible to a mind devoid of all imagination, and this extends to his opposition to Locke's separation out to subject and object, and Bacon's instrumentalism. For Blake's view is a quantum view; objects, like persons, both are and are not at a point *a*. Hence the ovum-larva-pupa-psyche that I have elucidated reappears in his *Milton* where "Single vision & Newton's sleep" arise from the undeveloped and threatening Ulro-hell. The caterpillar stage is the entire universe pictured as Blake's Mundane Shell (*M*, 19.21ff.), a merely embryonic world of reproductive life identified with Orc or Luvah. The chrysalitic world is sexual rather than truly human, the lower

Paradise, while the highest paradise is where humans are one with God. Although Peter Ackroyd attempts to reconcile Newton and Blake in his biography and almost succeeds,[12] in terms of modern sciences he has not quite made the case. In regard to at least some of them, especially quantum mechanics, science and art may in general terms remain incompatible. On the beach at Felpham, Ackroyd writes, Blake "saw 'particles bright' forming the shape of a man. Once more Newton's theory of particles had been given a mystical dimension."[13] The oneness into which the fourfold gathers allows the gates of perception to view all sides of an object simultaneously. Spatial and temporal barriers are overthrown as the poet is in various places at the same time. This enfolding allows multiple vision. The poet views the world with the spiritual eye beaming through the physical eye which becomes a "transparent eyeball," and transforms the universe into an emanation of soul.

NOTES

1. Abbreviations for Blake's works in the body of the text are:
   J: Jerusalem, Emanation of the Giant Albion
   M: Milton
   MHH: The Marriage of Heaven and Hell
   VFZ: Vala, or the Four Zoas
2. Cavelli-Sforza, 7.
3. Frye (1), 34.
4. Paley, 136.
5. Röstvig, 63.
6. Hilton, 16–17.
7. Sorensen, 36 n2, 25.
8. Hopper, 52.
9. Singhal, 1:155.
10. Frye, 368–69.
11. Abrahams, 2.
12. Ackroyd, 194.
13. Ackroyd, 219.

Part Two

Chapter Nine

From the Big Bang to Life on Earth

The fall of the universe into discrete matter from singularity is the real Fall over and beyond the biblical myth of the human Fall. It is that violent unraveling that confronts us today with the fate of planet earth. In its initial singularity the universe was enormously hot, far hotter than all the future-mortgaged fuel on earth could manufacture, but it existed in a state of perfect symmetry. As it expanded it cooled, and in each phase some of the symmetry relations between forces and particles disintegrated. So the cosmos fell from its pristine epoch consisting of unified forces and particles. From total unity, it collapsed into multiplicity. Time is transformed into space. Space itself is not smooth and continuous. The four fundamental forces then defined themselves: the strong and weak nuclear forces, electromagnetism and gravity. The first two forces bind quarks into protons and neutrons, and keep atomic nuclei together. Electromagnetism is crucial at the atomic level; it holds electrons to the nuclei, allowing atoms to coalesce into molecules. And gravity, which is caused by the warping of space and time (spacetime) by the presence of matter, dominates at the level of astronomical systems.

The Big Bang engendered a series of quaternal explosions. As the event unfolded, helium-3 picks up a further neutron to balance its two protons, and creates the catalyst for the formation of the earth in helium-4. The collision in turn of helium-4 nuclei gives rise to the unstable beryllium-8 through the triple-alpha process, and then one further interaction with helium-4 opens the way for the emergence of the carbon and oxygen crucial to life on earth. When carbon-12 is struck by another helium-4, oxygen results. The universe has moved beyond the Big Bang. Subsequently neon-20 loses its helium-4 to create oxygen, and both silicon and magnesium have their origins in reactions with the ubiquitous helium-4.

There can be no life without the 4 electrons found in the L shell of the carbon atom, giving carbon a valence of 4. This enables it to combine with a hydrogen atom to form the hydrocarbon molecule, methane (CH_4), which is one of the simplest organic molecules. As one scientist has written in regard to the anthropic principle:

"The laws of nuclear physics are just right for carbon nuclei to have formed in the interior of hot stars, later to be recycled into new stars and solar systems. And the laws of atomic physics allow for the formation of the most unlikely helical Tinkertoy molecules—the DNA that makes life possible. In other words, the pattern that seems to trump all other considerations is that the natural laws are conveniently fine-tuned to ensure our own existence. Physicists hate this idea. Especially string theorists."[1]

Before the formation of the quantum theory by Planck in 1900, and its application to the structure of atoms by Bohr in 1913, the nature of chemical bonds between two atoms could not be explained. It turned out to be the definite, quantitized symmetry that allows atoms to coalesce to form complex organic molecules. Carbon is of special significance because the number of electrons in its outer shell is just 4, which is half the number permitted in that shell. So carbon can absorb up to 4 electrons, and also lose up to 4 electrons.

During the Big Bang, in Population-II stars the nuclei whose atomic weights are multiples of 4 are favored because 4 is the atomic weight of He^4, which plays the dominant role in heavy-element build up. Population-I stars are formed from a chemical mixture that already contains heavy nuclei. Since these can capture protons in addition to He^4 nuclei, the restriction to nuclei whose atomic weights are multiples of 4 is finally removed. At this point, freedom in the sense of a certain randomness has replaced direct necessity. However certain further key numbers are thrown up when we consider the age of the universe, and its relation to the enablement of life on earth. Paul Davies has described this interconnection:

"The gravity acting between the constituents of an atom is some forty powers of ten (10^{40}) less powerful than the electrical attraction. . . . Asking now what is the longest natural unit of time available, we are led to the age of the universe, which has been calculated in various ways to be about fifteen billion years. In our fundamental subatomic units this turns out to be about 10^{40}, or one followed by forty zeros—the *same* enormous number by which gravity is weaker than electromagnetism.

The mystery is, why do we happen to be alive at just the epoch at which the age of the universe is equal to the magic number 10^{40}? Dirac argued that this number is so much bigger than those normally encountered in physical theory, like 4π and 12, that it is most unlikely that the above two ratios are equal by coincidence."[2]

Furthermore the total quantity of matter, based upon the atom, turns out to be the square of 10^{40}. These are the barest statistics of life, allowing us to function through the electromagnetic energy which is dependent on chemical bonding. Similarly the lifetime of a star is the ratio of the strengths of the gravity and electromagnetic forces which determine the functioning of the star— 10^{40}. So these subatomic units expressed to the power of multiples of 4 are the preconditions for our lives in the universe. Paul Dirac's variations on the Anthropic Principle have been summarized as

"Most physical and astrophysical dimensionless constants are of the order of magnitude of integral powers (positive and negative) of the number 10^{40}, where such numbers as m_p/m_e ~1,800 and hc/e^2 ~137 are said to be of the order of unity, the zero power of 10^{40}. He considered unlikely the accidental correspondence of the apparently unrelated, enormous numbers, and he suggested some unknown causal connexion."[3]

Then again, the 'coincidence' that creates the possibility of a total eclipse of the sun is related to the numerical conditions required for life on earth. The sun is 400 times bigger than the moon, but it is 400 times further away. As Guillermo Gonzalez says: "If we were a little nearer or farther from the sun, the earth would be too hot or too cold and so uninhabitable." Moreover we depend upon a large moon since its gravity ensures a relatively smooth elliptical rotation of the earth and prevents wild fluctuations in the climate. The moon started its trajectory away from earth several billions of years ago until it is close enough to the sun to cover it during an eclipse. This is a "timescale very similar to that of the appearance of intelligent life."[4]

On making his peace with mathematics in 1935, Einstein concluded that "the creative principle resides in mathematics. In a certain sense, therefore, I hold it true that pure thought can grasp reality, as the ancients dreamed."And the same year he quoted Leibniz approvingly to the effect that there is "a pre-established harmony between mathematics and the world of experience."[5] Newton had conceived the world as accommodated within absolute space and absolute time. Because in this system, absolute time was separate from absolute space, the Newtonian universe has $3+1$ rather than 4 dimensions. But with the space-time of relativity, space and time are so interwoven that the term '4-dimensional' is entirely accurate. This is because Einstein broke the Newtonian mirror when he said that different observers in uniform motion set up different systems of simultaneity. Einstein expressed it thus:

"It is a wide-spread error that the special theory of relativity is supposed to have, to a certain extent, first discovered, or at any rate, newly introduced the four-dimensionality of the physical continuum. This, of course, is not the case.

Classical mechanics, too, is based on the four-dimensional continuum of space and time. But in the four-dimensional continuum of classical physics, the sub-spaces with constant time value have no absolute reality, independent of the choice of the reference system. Because of this [fact], the four-dimensional con-tinuum falls naturally into a three-dimension and a one-dimension (time), so that the four-dimension point of view does not force itself upon one as *necessary*. The special theory of relativity, on the other hand, creates a formal dependence between the way in which the spatial co-ordinates, on the one hand, and the tem-poral co-ordinates, on the other, have to enter into the natural laws."[6]

In relation to the common coinage of everyday speech, this also has some-thing in common with the fifth dimension, which manifests itself through strings tangential to 'normal' reality and is made up of protons and other small particles. These strings then "move in more than the familiar four dimensions of everyday life—three spatial dimensions and one of time. Even though the gluons that make up the strings move in four dimensions, the string itself moves in five dimensions . . . The strings move in a five-dimensional curved space-time with a boundary. The boundary corresponds to the usual four di-mensions, and the fifth dimension describes the motion away from this bound-ary into the interior of the curved space-time. In this five-dimensional space-time, there is a strong gravitational field pulling objects away from the boundary, and as a result time flows more slowly far away from the boundary than close to it."[7] Some believe that this fifth dimension will be revealed in the near future through the Large Hadron Collider at CERN, the European Centre for particle physics. However Roger Penrose in his most recent profound med-itations on the problem remains convinced that such speculation offers no the-oretical advance whatsoever on the $1+3$ dimensions that we directly observe.[8]

In 1912 Apollinaire was to describe an artistic fourth dimension: "I would say that in the plastic arts the fourth dimension is generated by the three known dimensions: it represents the immensity of space eternalized in all directions at a given moment. It is space itself, or the dimension of infinity; it is what gives objects plasticity. . . . Greek art had a purely human conception of beauty. It took man as the measure of perfection. The art of the new painters takes the infinite universe as its ideal, and it is to the fourth dimension alone that we owe this new measure of perfection."[9] Charles Howard Hinton published his last in a set of fourth-dimensional fictions in 1904, and a number of Russian theo-reticians had surmised this dimension. Hinton wrote in that novel: "the human soul is a four-dimensional being, capable in itself of four dimensional move-ments, but in its experiences through the senses limited to three dimensions."[10] But Einstein's work, with its revelation of time as the fourth dimension, ended this fin-de-siècle spiritual speculation. Curved space destroyed the linearity of Euclidean geometry predominant since the Ancients.

Minkowski's theory of perduring objects, which opened the way for Einstein's advances, was based in absolute 4-dimensional space-time, replacing Newton's endurantist 3-dimensional Euclidean space. In this endurantist view, objects exist if they persist by being wholly present at different moments of time. But they do not have temporal thickness. On the perdurantist view, objects are actually 4-dimensional; they have extension in both space and time. A four-dimensional manifold of spacetime points includes all that happens in past, present and future. "Physical objects, on the perdurantist view, are represented in M [i.e. spacetime] diagrams by worldlines (worldworms) which could, in principle, be construed as strings (string bundles) of events stretched out in space-time."[11] This issue of the 3-dimensional and 4-dimensional persistence of object has become a crux in metaphysics.[12] The concept of endurantist ontology is that some object can be wholly present at all times they exist; the Minkowski universe has no absolute simultaneity.[13] In relativity, space (with co-ordinates x, y, z) and time (t) are welded to create 'space-time'. Two observers, moving relative to each other at constant velocity, will assign co-ordinates x, y, z and t to space-time in different ways. And under the laws of physics, they will agree upon the constant, c, the speed of light and the maximum velocity a particle can move. (João Magueijo has argued the speed of light was not a constant during the earliest evolution of the universe). In ordinary three-dimensional space, a 'vector' is a quantity with direction as well as magnitude. For example, velocities and electric field strengths are vectors. Given a cartesian co-ordinate system, a vector is represented by three 'components'. This generalizes to Minkowski space-time, and the corresponding type of quantity is a '4-vector'. It has four components. For example, the co-ordinates (x, y, z; t) of an event constitute a 4-vector. For a moving particle, the momentum and total energy make up another 4-vector. When Einstein spoke of semi-vectors, he put forward the principle that "the important point for us to observe is that all these constructions and the laws connecting them can be arrived at by the principle of looking for the mathematically simplest concepts and the links between them," because "nature is the realisation of the simplest conceivable mathematical ideas."[14]

There are a number of multiple-of-¼ relationships in the power laws in biology. The roots of this are in the branching networks of an organism. So metabolic rate varies with the ¾ power of the organism's mass—the bigger the creature, the slower its metabolism. Likewise, age of reproductive maturity varies with the ¾ power of mass. Life span, times of blood circulation, embryonic growth and development all vary with the ¼ power of mass. Rates of cellular metabolism, heartbeat, and maximum population growth all vary with the $-\frac{1}{4}$ power of mass. And so forth. Sizes of biological entities obey similar laws. Cross sections of tree trunks and of mammalian aortas both vary with the $\frac{3}{4}$ power of the mass. These relations are remarkably stable.

And, returning to the world of physics, in that concrete bunker between Cherbourg and Paris, an apparently entirely new form of nuclear matter has been formed. Tetraneutrons have been created, clusters of neutrons bound together in fours. It is hypothesized that tetraneutrons could be the nuclei of 'element zero,' an element without a proton currently missing from the periodic table. In some reactions where beryllium-14 breaks up to form beryllium-10 it is difficult to trace all four of the missing neutrons which just appear as a single flash. Four neutrons can somehow bind themselves together.[15]

The universe is some 15 billion years old and our own star, the earth, $4\frac{1}{2}$ billion years old. Signs of life on earth in the form of stromatolites (sediment cased in blue-green algae and bacteria) and cells can be traced as far back as 3.6 billion years. For over 2 billion years organisms were single celled (prokaryotic). The development of multicellular creatures (eukaryotic) were to require another 1.4 billion years of evolution. The 'big bang' in the organic world took place with the Cambrian explosion 542 million years ago, with a massive shaking down of phyla. And then life on dry land took a great leap forward with the development of insects around 400 million years ago in the Devonian age. Indeed the earliest creature, living out of the sea and with an insect's six legs, was recently found at the Rhynie chert of Scotland. It was millipede-like, *Pneumodesmus newmani*, technically a collembolan, and is some 428 million years old. The early insects were flightless, but by 325 million years ago they had developed wings. It is with the succeeding epoch of metamorphosis that this book will be concerned, a stage that has evolved over the past 250 million years.

It was Jan Swammerdam who discovered the transformatory process in nature, and this extraordinary entomologist and mystic was to be a major influence on William Blake. The two primary forms of insect transformation correspond to differing modes of human maturation. The earlier hemimetabolous form, which is gradual and incomplete, is favored by a minority of insects like the grasshoppers. This is historically the earliest manifestation of radical inner change, and the immature nymph is not unlike the completed imago, reaching that perfect state by a slow process of modification. This semi-metamorphosis involves the late development of genitalia, and wing buds turning into functioning wings only in the last moult to adulthood. On the other hand, holometabolous mutation comes relatively late in the geological record, and involves the separation of a larval and imago form, via the all-important intermediate stage—the pupa. So the caterpillar whose life consists of eating and fattening up passes through its pupal or chrysalitic form to become a butterfly — or beetle, bee or ant. These are the four fully metamorphic insects. Around 370 million years ago during the Devonian period, amphibians came onto land heralding the start of metamorphosis. Their progenitors had been the rhipidistians, a genus of fleshy-finned fish. Like today's amphibians, their eggs and tadpoles developed in the water but the adults could move on land. This pattern of growth would be paralleled when bees and moths appeared during the Cretaceous period some 140 million years ago, though it is estimated that insects had to have several goes before metamorphosis became established. Many modern Lepidoptera were present by the early Tertiary, 60–70 million years ago, and by 40 million years ago, all major butterfly families were present.

As James W. Truman and Lynn M. Riddiford discovered in 1999, the process of transformation is fired by the way a group of insect hormones, juvenile hormones and ecdysteroids interact during embryonic, larval and pupal stages. These juvenile hormones eventually disappear to become imaginal discs which program the adult.[16] Natural selection favors this sharp differentiation between caterpillar and adult. Vincent Wigglesworth explains: "the sort of body that will be best suited for chewing leaves or burrowing in the carcasses of animals will be very different from the sort of body required for flitting from flower to flower to seek a mate."[17] So the body grub specializes in eating, the imago in mobility, while the pupa is the powerhouse of the transformation. The division of labor, and of form, promotes the survival of the organism. Two different modes of maturation in insects exist. One where the larva or nymph phase passes over to the finished insect without an intervening stage of a pupa, and the other where the full four-set metamorphosis has to be undergone. This dichotomy has a profound significance throughout nature.

NOTES

1. Susskind, 36.
2. Davies (1), 171.
3. Dicke, 440.
4. Chown (2).
5. Pyenson, 153; Einstein (2), 136.
6. Einstein (3), 57–59.
7. Maldacena, 695.
8. Penrose, *passim*.
9. Apollinaire, 103–05.
10. Hinton (1), 22.
11. Balashov (1), 139.

12. Balashov (2), 644.
13. Balashov (1) and (2), *passim*
14. Einstein (2), 137, 136.
15. Bertulani and Zelevinsky, *passim.*
16. Truman, and Riddiford, *passim.*
17. Wigglesworth, 43.

Chapter Ten

Insects and Mind:
The Poetry of Dámaso Alonso

In defining the grounding in realism of Spanish poetry, Pedro Salinas drew a contrast between the *Cid* and the *Chanson de Roland*. There is, he said, an "absence in the Castilian poem of the unbridled imagination, incorrect proportions, scorn of the factual and the normal that abound in the French epic. Geography in the latter is imaginary. From time to time, fabulous beings, monsters, supernatural forces appear in the *Chanson*."[1] Dámaso Alonso, the most academic of the poets who attended the Góngora tercentenary in 1927, was to make the discovery in 1953 that there was a Spanish version of *Roland* that pre-dated the French version by several centuries. The Cid was "the first self-made man" in Spanish literature, and he was firmly based in the family, unlike the *Chanson*'s Roland, a being from another place. So when Alonso's Collection of poetry of two decades, *Sons of Wrath*, broke dramatically onto the post-Civil War literary scene in 1944, the verse has a traditional familiarity of tone and intimacy, almost a folk assurance.

Insects had appeared in one of his earliest poems, "The Poisoned Fly, or the Grand Ruse." Already the characteristic ease of voice is apparent:

"The fly scratches
its head.
A fit of nausea seizes it.
It lays down
on its side.
And gives a twitch (weary)."[2]

The simplicity is deceptive, for insects—their nature and place in the scheme of things—will come to dominate Alonso's poetic thinking to the point of self-declared obsession. Here though, already are indications of that philological interest that will develop into his full-scale academic treatises on the

73

Spanish language. In the first volume of his *Peninsular Linguistic Studies*, he remarks on the closeness of 'mosca', 'rasca', 'masca' and 'basca'—perhaps with an internal ear across half a century to this poem—and discusses their relation to the root '*oskra'.[3] Part of the fascination of the insect seems to be in its sounds, the range of its movements and noises that invade the life rhythms of humans:

> "There were tantrums, there was rage:
> to the floor it drops,
> and spins
> and raves
> buzzing, throbbing, twisting, snorting
> and it buzzes, how it buzzes, what buzzing, such buzzing!"[4]

The use of the dialect word 'runflando' ('snorting') of Santander derivation reminds us that Alonso was, by birth, from the north of Spain, though from Galicia rather than the Basque country. As in his compatriot, Rosalía de Castro, there is no formal partition between the poet and nature. The fly is almost beneath contempt, and yet there is a hint of metempsychosis about the way he addresses it: "It doesn't want to go. (To leave . . . how hard it is to leave)." And at the last, there is a moment of twinned vision as the insect causes him, as it does throughout his poetry, to consider the nature of life itself—

> "Oh, Lord of Life:
> but it's the 20th of May,
> but in reds and greens the garden erupts!"[5]

Is it amusing, banal or genuinely disturbing that the fly is now reduced to a spreading, cloudy, invasive stain? Always with a strain of irony, Alonso will return to such reflections in a poem of 11 years later.

"The Insects"is the first poem actually written that appears in *Sons of Wrath*, dating from 1932. The poem comes with the apparatus of a Preliminary Note, a wonderful serio-comic burlesque on Eliotian pretensions in the form of a letter to a fictitious correspondent. The insects had broken in on the poet as he worked, attracted by the light, a physical light to the writer's intellectual lamp. The teeming life is tropical:

> "And above the lamp, above my head, above the table, hurtled
> huge bands of insects, some sticky and soft, others with a
> fleeting flash of stone or metal: brilliant, tough, lumbering
> coleopters; miniscule restless hemiptera, and others rose from
> the ground flying silently, with their sweet bug-odour;
> monstrous, grotesque orthoptera; lepidoptera in miniature."[6]

Does this profligacy in life and death signal the destruction of the human, the insects impervious and liberated survivors through the millennia, able to surmount the nuclear? Or can these alien creatures be woven into a cosmic vision, perhaps a comic vision, an aesthetic? Despite the magnificent tracery on the wings, the note decides that these insects are nothing other than "the most beautiful cesspools."[7] However critics have missed the extremely serious note behind this travesty of an academic Preliminary Note: the hallucination brought on by the nocturnal invasion is also a crisis of ethical judgment, a dilemma despite the Lilliputian realm the poet has entered.

A sense of putrefaction is present in the first lines of *Children* "Madrid is a city of more than a million corpses (according to the latest statistics)."[8] Much cerebral fluid has been spent arguing whether Alonso belongs to the anti-Franco social poets of the 1940's and 1950's, but there seems from his comments little doubt that he could be viewed as their immediate precursor, even though he simultaneously viewed the decay in Madrid as transcendental. As he himself wrote, the book is "a universal, cosmic protest that clearly includes all those other piecemeal angers."[9] He is overwhelmed by the creatures—

> "The insects were causing me an extraordinary amount of bother
> because there's no doubt I don't trust insects,
> so much attention, so many legs, heads, and those eyes
> oh, above all those eyes
> that don't let me control the dread of night,
> the terrible dryness of nights when insects are buzzing about."[10]

The profligacy of life causes him to arraign the insects, and try to explain the interweave of "the world of my flesh (and the flesh of the insects)."[11] In this poem at least, which is an early poem, he does not fully grapple with bioethical problems, but ends with the curse—"the bloody insects."[12]

As his prefatorial note had indicated, the poet is moving toward the gothic, defining insect odour as "multiplying their frigid animosity, their anti-human malice, their power to wound or sting in the pituitary and cause a shock, a frenetic alarm in some nerve centre."[13] Alonso makes this even more manifest in a note written for his *Selected Poems* of 1969: "The author as a participant in life loves it intensely; hates at the same time the monstrous injustice that prevails in life. A consequence of this is to consider as monstrous the totality of life. But already in this sense the word 'monstrous' acquires another value: life is monstrous because it is inexplicable. Each creature is a monster because it is inexplicable, alien, absurd. (It is the primary meaning of *monstruum* in Latin)."[14] And so in the poem "Freaks," Alonso begs for illumination to overcome the terror that gnaws at him, and of which the insects are the objective proof. Indeed the poet himself becomes implicated as "the frenetic Dámaso," who is as "horrible" as "this yellow centipede that beseeches you with all its crazy tentacles."[15] Proximity to the abundance of nature is translated into a very particular sense of identity, not so much a concept of humility, and only passingly what Andrew Debicki sees as the necessity for the poet to "abandon his self-centeredness."[16] Rather it is an attempt to get the focus just right, to size up humanity's place in the dynamism and proliferation of an earth that can foster such insect life, and then to re-fit our species into its true place in nature. Expressed ideologically, it is another aspect of recovering from imperial illusions and privileging, which has distorted our way of understanding our position in the cosmos. Alonso has bravely, and early, set out to re-define human limits.

Like "The Insects," "Elegy for a Blue Hornet" has a preface, albeit brief setting the poem in media res. There is to be nothing parnassian about this poem, and indeed throughout his criticism Alonso employs the word 'parnasiano' in a derogatory sense, undoubtedly partly as a result of the influence of Gerard Manley Hopkins's aversion to the parnassian tradition. Indeed he translates six of Hopkins's poems.[17] The foreword begins—"Yes, I murdered you stupidly. Your buzzing irritated me while I was composing a beautiful, sweet love sonnet. And it was a consonant for *-úcar*, to rhyme with *azúcar* (sugar) that I was lacking. *Mais qui dira les torts de la rime?*"[18] Ending with the allusion to Verlaine, poet of the frontier between dream and wakefulness, it is a parody of a literary note and brings to mind Sterne's Uncle Toby who magnanimously restrained himself from reprisals on an unwary fly—"go poor Devil, get thee gone, why should I hurt thee?—This world surely is wide enough to hold both thee and me."[19] However, serious matters of nature's intentions are at stake for Alonso. The poem darts immediately into the moment of the poet's emotional recoil:

> "Then I felt anguish
> and looked at you closely: you're very handsome."[20]

Studying the hornet, he starts to consider its animal soul by way of its great eyes, and concludes that such a soul must be like a great fire, a furnace of colours, like a lighthouse lantern. This contemplation takes him back in memory to his childhood Galician house, far from his house in the outskirts of Madrid. Focus sharpens as the lengthy lines of ponderings give way to:

> "When I killed you,
> you were looking outside
> at my garden. This clear December
> presses on me its colours and light . . ."

But the scene has ossified, petrified into "blocks of marble, brutally" and poet is reduced to cataloguing what makes up the world, "lights and forms,/tree, bush, flower, hill, sky." The invasion of nouns by verbs then emphasizes the teeming activity which waits to overrun the fixities of the definite, "swarming" and "a sweet ferment no more."[21] The mention of "sweet" reminds us of the rhyme the poet was seeking with "azúcar" before the hornet interrupted, and also that some of the values he has held dearest are crumbling in this blitz of nature, not only moral values but also that of the sonnet form itself, as indeed Alonso himself remarks.[22]

Now the poet has revised his place in the scheme of things: "Oh poor being, equal, equal, you and I!" Its head is now "noble," and he notices that in death its proboscis has remained erect and extended. And then "azúcar" reappears, but now not in the context of the love poem as originally conceived so much as a genuine wondering at "What juices or what sugars/were you/ voluptuously/breathing."[23] This sentence as a whole which partly constitutes a question, ends with an ironical comparison of the death of an insect with the sinking of a transatlantic liner, tumbling from the poet as he himself falls into the abyss of the ethical questions unleashed by his cogitations. Much of the humour arises from the sheer apparatus that the scholar-poet brings to bear on such an apparently miniscule problem, which almost seems to be "nada," a word that occurs on four separate occasions at the centre of the poem. Alonso pointed to D.H. Lawrence's "Snake" as a touchstone for this poem. As one might expect from the very nature of snakes, the Lawrence poem is more strongly physical, the presence of snake and man more physically solid, the relationship involving more substantial give and take. But there is a similar fear of what is perceived as a telluric nothingness:

> "And as he put his head into that dreadful hole,
> And as he slowly drew up, snake-easing his shoulders,
>   and entered farther,
> A sort of horror, a sort of protest against his withdrawing
>   into that horrid black hole,
> Deliberately going into that blackness, and slowly drawing
>   himself after,
> Overcame me now his back was turned.
> I looked round, I put down my pitcher,
> I picked up a clumsy log
> And threw it at the water-trough with a clatter."[24]

Lawrence can remain relatively calm, since the snake has only invaded outdoor territory. But Alonso's insects refuse to occupy their own kingdom. They not only invade the poet's house, but even more his mind and soul. They

worry, chivvy and disturb, bothering him to his nerve ends. He needs to
gather all his intellectual wits to assess his connection to this now dead hornet. The English writer, though, is more aware of his mind as something
alienated: "I despised myself and the voices of my accursed human education."[25] In a sense, Alonso's "Moscardón" for all its irony and pathos—the
human overwhelming the microcosmic with his rationalising—is a far more
serious poem than "Snake." Indeed Jon Silkin has suggesting in regard to the
Lawrence poem—"Not that fear isn't potent, but on the evidence of the *poem*,
there is only a modicum of it."[26]

A comparison with Lawrence's "Butterfly" is, perhaps, even more instruc-
tive. Admittedly the creature concerned is more attractive than Alonso's hornet, but given its fragility, it is remarkable how paradoxically solid this Large
White becomes: "Butterfly, why do you settle on my shoe, and sip the dirt on
my shoe,/Lifting your veined wings, lifting them? big white butterfly!"[27] The
Spanish poet in contrast is full of anguish, and despite the various details of
the hornet that he gives, more abstract than Lawrence. For Alonso has raised
the encounter to a metaphysical level:

> "Two, three times
> an obstinate segment
> quivered in the air, as if a cryptic epitome
> of the throbbing world, its final
> message.
> And you were a thing: a corpse."[28]

The poem shudders to a halt in this penultimate strophe to define the organism as a thing, reified by the poet's thoughtless act, and calling up again the
reification of the December scene earlier in the poem. The poet is left ex-
claiming: "Oh could I only/ have given you life again,/ I who gave you
death!" reduced as he is to the tentative abstractions of the subjunctive.[29]

In the first edition of *Hijos de la ira*, "La Obsesión" concluded the book,
and it returns to what Alvarado de Ricord terms "the fruitless search."[30] Some
of his intense pessimism is probably generated by the situation in Madrid in
the 1940s, but also by the legacy of his early reading of Darío, and his af-
fecting of a translated European melancholy. As the poem's title indicates, the
insect question has become an obsession, and a certain weariness informs its
opening lines addressed, familiarly this time, to "a green hornet": "You. Always you. There you are."[31] The insect has taken on something of the de-
monic, at the least of the impish, and also perhaps he has recognised the
species as symbol of his own poetic daemon, a point at which inner and outer
cannot be distinguished. A profound depression has descended on him, traceable to a line from the "Elegía a un Moscardón Azul" where he spoke of "that

necessity of projecting futures/ that we call life)."[32] Written between 1944 and 1945, and absent from the first edition of *Hijos*, "La Obsesión" expresses the mood of a poet living in a society where, compared to the years of the Republic from 1931 to 1939, the normal processes of direct statement are suspended. The hallucinatory quality of the poem and the despair -"No, no more, no more, make an end, make an end"[33]—seem not only metaphysical but implicitly political as well. The poet is writing in a vacuum in which his fellow readers from the great Góngora gathering of 1927 are dispersed or dead. And again, there is an echo here of Hopkins—of his "No worst there is none" and "O then, weary then why should we tread?"—a poet whom Alonso characterised as "the creator of an intense and obsessive poetic world, with an emotion communicable quite beyond the word."[34] And so Alonso envisages the time when he is consumed by "the high, pure flames" and "the final victory" can be proclaimed "by human reason that has extinguished itself."[35] The concluding poem of this volume hypothesizes a poet who has sprouted wings through the love of a woman. Here at last the poet discovers that "beneath the shoulders were opening/ two wings,/ strong, immense, of an immortal white."[36]

Alonso returns to the world of the insect in his last poems where he begs that his sense of sight be preserved, though by now the concept of a triumphant post-rational creator-poet has receded. He prays, with characteritic burlesque modified by an undertow of high seriousness, that he be not left like a "sad earthworm, a muddy bereaved larva."[37] Nor

> "like a gigantic slug, like a slow slug
> the size of fifty elephants,
> a ghastly roundback black-green, or of the saddest
> khaki colour . . ."[38]

His entomological intimations remain perfectly intact throughout a critical career that saw a number of tactical advances and retreats, from his expositions and championing of Góngora to later reservations about the Golden Age poet (similar to Lorca's briefer oscillation); and from emphases on structure and semiotics to a purer, and simpler approach to literature which he thought, for example, more appropriate as an approach to Hopkins than the more elaborate externality of the English critics.[39] In the end, Alonso's poetic deliberations can be best interpreted as in the tradition of San Juan de la Cruz's wrestlings with the "dark night of the soul." Indeed as he said in his Address, "Scylla and Charybdis in Spanish Literature," at the Ateneo de Sevilla during the historic 1927 discussions, the twentieth century Hispanic Renaissance is a synthesis of the Spanish medieval tradition and the European Renaissance.[40]

NOTES

1. Salinas, 19
2. Alonso (2), 171.
3. Alonso (4), 490.
4. Alonso (2), 172.
5. Alonso (2), 172.
6. Alonso (3), 142-43.
7. Alonso (3), 144.
8. Alonso (3), 73.
9. Alonso (5), 194
10. Alonso (3), 145
11. Alonso (3), 146.
12. Alonso (3), 147.
13. Alonso (3), 144.
14. Alonso (5), 194.
15. Alonso (3), 119.
16. Debicki, 63.
17. So, for example, he writes of Aleixandre: "Sometimes verses result that, by contrast, seemed almost parnassian were it not for the internal irrepressible energy." (Alonso (6), 382). The Hopkins translations are also to be found in his *Poetas Españoles*, 393-97. The poems are: "Pied Beauty," "The Starlight Night," "Hurrahing in Harvest," "The Leaden Echo and the Golden Echo," "Carrion Comfort," and "No Worst there is none."
18. Alonso (3), 113.
19. Laurence Sterne, *The Life and Opinions of Tristram Shandy, Gentleman* (Gainesville: University of Florida, 1978), 1:131.
20. Alonso (3), 113.
21. Alonso (3), 114.
22. Alonso (5), 197.
23. Alonso (3), 115.
24. Pinto (2), 1:351.
25. Pinto (2), 1:351.
26. Silkin, 208.
27. Pinto (2), 2:696.
28. Alonso (3), 116.
29. Alonso (3), 117.
30. de Ricord, 122.
31. Alonso (3), 131.
32. Alonso (3), 115.
33. Alonso (3), 133.
34. Gardner, 100, 93.
35. Alonso (3), 133.

36. Alonso (3), 173.
37. Alonso (2), 25
38. Alonso (2), 49.
39. Alonso (6), 388, 391.
40. Alonso (1), 23.

Chapter Eleven

A.R. Wallace's Swallowtails: Mimicry and Evolution

Alfred Russel Wallace had a recurrent dream during his early childhood in the Usk Valley "as of some creature with huge wings." He ultimately traced its origin to a funeral escutcheon he had seen, and which had played on his imagination like "an unmeaning jumble of strange dragon-like forms."[1] Prophetic that his most memorable experiences as a naturalist should involve beasts with huge wings! For central to Wallace's life would be the insects, and specifically the tropical Lepidoptera, butterflies and moths. The structure and coloration of Lepidoptera play a major role in his disputes with Darwin. The thick scale exterior from which the species get its name (*lepis* = scale) evolved from ancestral hair cover. Indeed moths and the more primitive butterflies (*Hesperidae* or Skippers) retain dense hair cover at the thorax. The selective advantage of scales for the diurnal moths and butterflies arose from their allowing a cooling cover in the sun, and protection from water loss. Moreover, the bright coloration of scales in certain groups afforded both protection and a signalling ability to find for reproduction partners.

Amabel Williams-Ellis's book title christens Wallace 'Darwin's Moon,' yet the range of his concerns and expeditions outstrip Darwin's. He visited the Malay Archipelago, Singapore and the Amazon on insect-gathering expeditions for his financial livelihood, while he gave his name to the Wallace Line dividing Australian from Oriental flora and fauna, which with modifications still holds. As a collector, he specialized in Pieridae, whites and yellows, compiling a catalogue of all known in the Indian and Australian regions, and describing 50 new species. Of the Cetoniidae, he tabled 70 new species. This hugely variegated and even aesthetic fieldwork among "fresh and living beauty" as he called it, surely had its impact on the crucial issues that were to drive a serious theoretical wedge between Darwin and himself.

He gave his name to a number of Swallowtails: *Papilio polydorus leodamas* Wallace (now *Pachliopta polydorus leodamas* Wallace); *P. lorquiniamus phillipus* Wallace; *P. paradoxa aenigma* Wallace; *P. pericles* Wallace; *P. albinus albinus* Wallace; *Troides oblongomaculatus papuensis* Wallace; *T. plato* Wallace; *T. criton celebensis* Wallace; *Graphium codrus gilolensis* Wallace; *G. wallacei, Trogonoptera brookiana brookiana* Wallace. The swallowtails are so called as most have tailed wings. They subdivide into three main groups: the true swallowtails (*Papilio*), Kite swallowtails (*Euritydes* and *Graphium*) which have narrow pointed wings like an old-fashioned kite, and the poison-eaters (*Parides* and *Troides*) whose caterpillars feed on *Aristolochia* vines from which they derive their poisonous substances, and who are therefore mimicked by some non-poisonous butterflies for protection. (There are also the Apollos and one primitive swallowtail found only in Mexico, *Baronia brevicornis*.) Lepidoptera have contributed greatly to the understanding of the mechanisms of evolution because of the clear examples of sophisticated protective strategy which gives them a role in natural selection. In Britain and the United States, there is the false eye of the Emperor Moth that on open moorland gives even the human observer a shock, or the ragged edges of the Comma butterfly which works in with the scalloped leaves of a tree.

On 23 February 1867, Darwin wrote to his younger colleague:

> "On Monday evening I called on [H.W.] Bates, and put a difficulty before him which he could not answer, and, as on some former occasion, his first suggestion was, 'You had better ask Wallace.' My difficulty is, Why are caterpillars sometimes so beautifully and artistically coloured? Seeing that many are coloured to escape dangers, I can hardly attribute their bright colour in other cases to mere physical conditions."[2]

Wallace was at the time preparing "my rather elaborate paper" on 'Mimicry and Protective Colouring' for the *Westminster Review*.[3] Mimicry shields some organisms that are palatable to predators, by their evolving a similarity to those unpalatable. Wallace explained that a caterpillar's bright colours, by making it so visible, encouraged its enemies to distinguish it at a glance from others attractive to taste. (It has also recently been argued that there are palatable lepidoptera that imitate the movement behaviour of unpalatable prey.)[4] The principle of ostentatious marking holds especially with butterflies of the Papilionidae and Heliconiidae families, which have a strong smell and unpleasant taste protecting them from insectivorous birds and other creatures.

When Darwin objected on the grounds of differences in coloration between male and female, Wallace replied drawing on his observations of the lesser mobility and vulnerability of the female during ovipositing:

> "Your objection that the same protection would to a certain extent be useful to the male, seems to me utterly unsound, and directly opposed to your doctrine so convincingly urged in 'Origin' *that Natural Selection never can improve an animal beyond its needs.*
>
>   A male, being by structure and habits less exposed to danger and less requiring protection than the female, cannot have more protection given to it by Natural Selection, but a female must have some extra protection to balance the greater danger, and she rapidly acquires it one way or another."[5]

Already in a paper of 1865, Wallace had begun to dispute the territory which Darwin put outside the laws of natural selection. He shows there how the more subtle camouflage of female butterflies is occasioned by "their slower flight laden with eggs."[6]

Mimicry was becoming a key issue with evolutionists even as Darwin published his *The Origin of Species* in 1859. Wallace declared that Darwinian principle "leads us to seek an adaptive . . . purpose . . . in minutiae which we should otherwise be almost sure to pass over as insignificant or unimportant."[7] Wallace pointed out that the two sexes of *Papilio paradoxa* exactly imitate *Euploea midamus chloe*, and appear in the same districts. So similar are they that he "could hardly ever distinguish them on the wing." The male of the *Euploea* "has the fore wings of a brilliant metallic blue, with faint bluish-white spots, while the hind wings are uniform brownish black. The female differs considerably, the hind wings being covered with narrow white lines radiating from the body, and having a marginal row of white spots."[8] We should hardly be surprised that mimicry plays a role in organic nature. After all it is common in the world of physical matter. Every newly created particle is accompanied by a sort of 'negative image' partner, a so-called antiparticle somewhere in the universe however distant. So an electron (which carries a negative electric charge) is always created along with an antielectron or positron, which has the same mass as the electron but an opposite, positive, charge. Every proton is shadowed by an antiproton. This doubling process is a feature of nature, and is imaginatively built upon by many writers from Hogg to Nabokov.

The arguments between Darwin and Wallace became sharper with time, as Darwin dug his heels in, insisting that the brighter wings of males, whether bird or butterfly, was primarily as an attractor to females. Since butterflies respond to colour when courting, sexual selection resists colour change in males. Hence the male swallowtail retains an ancestral colour pattern. When in *The Descent of Man*, Darwin writes that "the colours of caterpillars are mostly protective, being due to natural selection alone, while those of butterflies are mostly attractive, being largely due to sexual selection," Wallace answers in 1879 that this is a slurring over of "what is really a stupendous difficulty in the way of the theory. So far from the colours of caterpillars being 'mostly protective' every entomologist knows that a large number of caterpillars in every part of the world are conspicuously coloured, and what is more to the point that their colours are as brilliant and varied as those of butterflies themselves, if we take into account the nature of their integument, the small amount of surface, and the uniform cylindrical form of their bodies."[9] What Darwin had called 'minor' causes of modification—immediate conditions, use and misuse, variability and habit together with more sophisticated secondary traits of beauty—had for Wallace to come within the purview of the necessities of natural selection. Martin Fichman puts the matter succinctly:

> "To Wallace, evolutionary explanations of behavioral or physical traits predicated on an aesthetic sense in lower animals were unacceptable. The discontinuity between human higher faculties and the mental processes of the rest of the animal kingdom had become axiomatic for him. Wallace declared that imputing aesthetic tastes to birds (and insects) was an anthropomorphism as unwarranted as that made by 'writers who held that the bee was a good mathematician, and that the honeycomb was constructed throughout to satisfy its refined mathematical' sense."[10]

Wallace was already becoming what he later christened himself: "more Darwinian than Darwin," which is quite strange since he had such very serious cultural and spiritual differences with the author of *Origin of Species*.

There is no mimicry among butterflies in Britain, though it does occasionally occur among moths, as with the Broad-bordered Bee-hawk and Hornet Clearwing. (The lines of Robert Browning, which especially appealed to Nabokov, play on this lepidopteral feature—"lichens mock/ The marks on a moth, and small ferns fit/Their teeth to the polished block.") Abroad though there are the Papilionidae considered, as Wallace put it, "by all the older writers to be the princes of the whole lepidopterous order," and "with opalescent hues, unsurpassed by the rarest gems."[11] Above all for the student of evolution, "they exhibit, in a remarkable degree, almost every kind of variation, as well as some of the most beautiful examples of polymorphism and of mimicry," along with a wide variety of localization.[12] As Wallace remarks in his article on "The Malayan Papilionidae or Swallow-Tailed Butterflies, as illustrative of the Theory of Natural Selection," these insects offer "immense development and peculiar structure of the wings which not only vary in form more than those of other insects, but offer on both surfaces an endless variety of pattern, colouring, and texture."[13] Moreover the prevalence of mimicry was extensive among the swallowtails: "In all parts of the world there are certain insects which, from a disagreeable smell or taste, are rarely attacked or devoured by enemies. Such groups are said to be 'protected,' and they always have distinctive and conspicuous colours."[14] Their mimics can be of either sex, "but most frequently it is the female only that is thus modified, especially when she lays her eggs on low-growing plants; while the male, whose flight is stronger and can take care of himself, does not possess it, and is often so different from his mate as to have been considered a distinct species."[15]

It was Wallace, rather than H.W. Bates (or Fritz Müller), who defined the pre-eminence of colour's role as the key to warning predators, rather than smell or any other characteristic. So in his 1866 paper on "Natural Selection," he writes:

"1. In all cases of mimicry, the resemblance of the one species to another in a different group is entirely superficial, and is always strictly confined to those characters which cause the one to *look like the other.* [This is confirmed by work on mithochondic evidence on mimic butterflies and geography] The structure, the habits, the form of inconspicuous parts, the nature of the food, or the character of the larva and pupa, are not, as far as we know, ever modified in a similar manner. . . .
2. There are no grounds for believing that minute details of colouration and marking are due to climatal conditions at all, still less that they can be produced so identically alike in species of groups widely differing in organization. . . ."[16]

A section of *Malay Archipelago* gives another angle on Wallace's observations on mimicry. Once again, they are very precise, and emphasize his intimation that the Lepidoptera have a special role in the theory of evolution. The handsome *Papilio memnon* is "a splendid butterfly of a deep black colour, dotted over with lines and groups of scales of a clear ashy blue" with wings "five inches in expanse, and the hind wings are rounded, with scalloped edges. This applies to the males; but the females are very different and vary so much that they were once supposed to form several distinct species. They may be divided into two groups—those which resemble the male in shape, and those which differ entirely from him in the outline of the wings." One of the females led Wallace "to discover that this extraordinary female closely resembles (when flying) another butterfly of the same genus but of a different group (*Papilio coön*)." This resemblance occurs because "the butterflies imitated belong to a section of the genus Papilio which from some cause or other are not attacked by birds, and by so closely resembling these in form and colour the female of Memnon and its ally, also escape persecution."[17]

Dawkins asserted there is a supergene that effects the approximation of the imitator to the model butterfly.[18] This does not now seem to be the case. Instead in the course of time the approximation becomes perfect resemblance. This occurs as a gradual process of natural selection; an imitation that becomes closest in appearance to the unpalatable organism has a greater chance of escaping the predators, and this imitation over generations becomes the norm by way of genetic drift. The biochemical process is created by a morphogenic diffusion that stimulates a gene or colour-specific enzyme to generate a stable spatial pattern. Or in another type of mimicry where a butterfly or moth may resemble a leaf, "Even the peculiar colours of many animals, especially insects, so closely resembling the soil or the leaves or the trunks on which they habitually reside, are explained on the same principle [as the giraffe's neck]; for though in the course of ages varieties of many tints may have occurred, *yet those races having colours best adapted to concealment from their enemies would inevitably survive the longest.*"[19] In "On the Law Which Has Regulated the Introduction of New Species" of 1855, Wallace concluded that "every species has come into existence coincident both in space and time with a pre-existing closely allied species." So new species arise through variations which enhance survival: "The superior variety would then alone remain, and on a return to favourable circumstances would rapidly increase in numbers and occupy the place of the extinct species and variety. The *variety* would now have replaced the *species*, of which it would be a more perfectly developed and more highly organized form."[20]

It is infamous now that, despite its title, *The Origin of Species* does not centrally address the problem of speciation. As Wilma George put the matter: "It was concerned with establishing natural selection as the directive force in the steady evolution of one species into another over geological time. Little was said about the circumstances in which a population could be divided into two or more groups each of which would become a species contemporaneously."[21] Wallace spelt out the conditions under which speciation to take place:

"(1) That whatever the amount of variability of a species, no general modification of it will occur so long as the environment remains unchanged; and (2) that when a permanent change (not a mere temporary fluctuation) of the environment occurs—whether of climate, of extension or elevation of land, of diminished food-supply, or of new competitors, or of new enemies—then, and then only, will various specific forms become modified, *so as to adapt them more completely to the new conditions of existence.* . . .

"In order to be developed through natural selection a particular variation must not only be *useful*, but must, at least occasionally, be of such importance as to lead to the saving of life, or to use Professor Lloyd-Morgan's suggestive term, be of 'survival-value.'"[22]

Speciation remains a hotly contested issue. It was raised by E.B. Poulton in "What is a species?" His species concept elucidated in that 1904 address is, as James Mallet explains, that they are "*syngamic* (i.e. formed reproductive communities), the individual members of which were united by *synepigony* (common descent). Poulton's species concept was informed by his knowledge of polymorphic mimicry in *Papilio* butterflies: distinct non-mimetic male and mimetic female forms were members of the same species because they formed *syngamic* communities."[23] Significantly Wallace had just given Poulton a book on mimicry in December 1903, which included the first mimicry papers, by himself, H.W. Bates and Roland Trimen.

So the study of mimicry is at the heart of evolutionary theory, and it is again interesting that the world of insects in which Wallace had immersed himself from an early age provides the key to these issues. A core of the argument centres on structure, and by implication even upon the nature of language. In overthrowing Paley's argument for the existence of God—which might agnostically be read as Perfection—from design, Darwin simultaneously threw out most of the aesthetic sense, the baby with the bathwater, except in the instance of sexual selection of mates. Significantly Wallace's much-quoted rhapsodic response to *Ornithoptera croesus* had already suggested an almost aesthetic function for lepidoptera. When Darwin writes in his Autobiography, "we can no longer argue that, for instance, the beautiful hinge of a bivalve shell must have been made by an intelligent being," the premise is rightly the fixity of the laws of nature.[24] But in this epoch of a far greater knowledge of internal mechanisms, it is now possible to unify structural design and nature's systematizations. This is not to return to the Lamarckian belief in the existence within organisms of a built-in drive toward perfection as part of the heritability of acquired characteristics. Darwin's discoveries on speciation rejected the latter and, as the neo-Darwinians insist, speciation is usually the consequence of the divergence of populations separated by a geographical divide. (It was Ernst Mayr who was to formulate the species concept based on reproductive isolation). This is also the difference between Lamarck's vertical and Darwin's horizontal time; the latter is concerned with the origin of the diversification of species, and dismisses Lamarck's model of groups of organisms progressing independently by spontaneous generation towards perfection. So selection is the outcome of variation, and mistake rather than deliberate design is the pivot. During DNA copying sequences in reproduction, errors enter. The resultant copy is imperfect, and the proteins coded differ. Evolution occurs through such discrepancies. Far from striving for perfection, the root of the variations from which natural selection produces evolutionary change is such error in genetic transmission. So it is the concept of process that reunifies design and fixed laws.

In regard to further differences between Darwin and Wallace, A.J. Nicholson has defined a variation of emphasis between Darwin's 'competitive selection' and Wallace's 'environmental selection.' Under competitive selection, "the individuals that do not survive are not really unfit at all. They are simply *less fit* than the survivors and are eliminated only because of the presence and preservation of the more fit individuals."[25] For Wallace though it was the external environment that set an absolute challenge to be met in order to survive so that, as he expressed it, "if there were some alteration in the physical conditions" such as an infestation of locusts extinguishing a parent species, "the *variety* would now have replaced the *species*, of which it would be a more perfectly developed and more highly organized form."[26] In time the variety would become the new species. At the same time, Darwin was aware of both factors even though he continually turned to the arguably more societal argument. Significantly, Wallace had no interest in artificial selection and domestication, and always took his examples from natural populations. This, however, meant he failed to explain the divergence of species. "Every species comes into existence coincident in time and space with a preexisting closely allied species," he wrote in *Annals and Magazine of Natural History* through which he first came to Darwin's notice. So he described the fact but not the mechanism of divergence. Or as Peter Bowler has it, "Wallace simply assumed that species split into varieties—he did not seek to explain how this all-important first step occurs."[27] The process could only be explained by those societal causes that Wallace shunned, for it is diversity which reduces competition between organisms until such a point as divergence of species occurs. Perhaps the best definition of speciation remains that of Karl Jordan produced in 1896 while employed by Walter Rothschild, and favoured by Mallet:

"A species is a group of individuals which is differentiated from all other contemporary groups by one or more characters, and of which the descendants which are fully qualified for propagation form again under all conditions of life one or more groups of individuals differentiated from the descendants of all other groups by one or more characters."[28]

But on a basic question such as the reason for the upright posture of humans, Wallace was able to provide the environmental answer that was lacking in Darwin. His adaptionism undoubtedly led him to over-estimate utility as an all-pervading factor, but at the same time it did mean that he could not accept the relativism with occasional lapses into fundamentalism in regard to artistic and spiritual matters, allowed by the pluralism of Darwin.

It was in 1869 when Wallace reviewed the tenth edition of Lyell's *Principles of Geology* that for the first time he stated publicly and categorically his belief that natural selection could not account for the growth of the mind of humans. Although it "may teach us how, by chemical, electrical, or higher laws, the organized body can be built up," yet it "cannot be conceived as endowing the newly arranged atoms with consciousness."[29] Even as early as 1864—the date of *The Origin of Species* was 1859—Wallace had insisted that the whole of natural selection had changed with the evolution of the human brain. But Darwin wrote in one of his *Transmutation Notebooks* enquiring "why is thought being a secretion of the brain, more wonderful than gravity a property of matter? It is our arrogance, our admiration of ourselves."[30] What would he have made of Gerard Manley Hopkins's description of Purcell's music as having "uttered in notes the very make and species of man as created both in him and in all men generally"?[31] Darwin's Autobiography attests to an outlook that has led to many of the cul-de-sacs in contemporary thinking: "My mind seems to have become a kind of machine for grinding general laws out of large collections of facts, but why this should have caused the atrophy of that part of the brain alone, on which the higher tastes depend, I cannot conceive."[32] This Gradgrindery was the root of Darwin's later inability to react to music and poetry, other than his beloved Milton, though novels being a product of "not of a very high order" of imagination were read for relaxation.[33] The primary discoverer of the laws of evolution himself becomes crucial evidence in favour of Wallace's reservations as to the universality of their application. In the course of his long life, Wallace came to feel the need for a more comprehensive view of the human situation, of humanity's relation with nature, than evolutionary theory allowed. The spiritualism he came to later in life was part of that. Certainly Wallace evinces a deeper feeling for nature than Darwin, not least in his response to the variety of lepidoptera. Again, the Papilionidae which include most of the great Birdwings, offered much for the theoretician as well as the naturalist and even aesthete: "the family presents us with examples of difference of size, form, and colour, characteristic of certain localities, which are among the most singular and mysterious phenomena known to naturalists."[34] And when he saw the *Ornithoptera poseidon*, he experienced "the joy which every discovery of a new form of life gives to the lover of nature"—"It is one thing to see such beauty in a cabinet, and quite another to feel it between one's fingers, and to gaze upon its fresh and living beauty, a bright-green gem shining out amid the silent gloom of a dark and tangled forest."[35]

As we have seen, Wallace maintained that the human brain specializes in "the evasion of specialization."[36] He viewed this evasiveness as characteristic of the higher evolution of humans. The brain, after all, is an intricate interrelation between over 100 billion cells and even at the merely physiological level, the molecular events accompanying thought and its attendant, memory, are hardly conducive to the banal fixities of one-sidedness. So Sir John Eccles asks whether there is "some process that we could call *genetic dynamism* whereby the hominid brain inevitably develops further and further beyond natural selection?"[37] The configurations of both natural and artistic laws in their spontaneous symmetry strike us as beautiful, and present themselves today with the urgency Wallace expressed over a century ago when writing of the Bird-winged tropical butterflies and the Great Bird-of-Paradise. Humanity, he warned, threatened so to disturb the balance of nature as to cause the extinction of those creatures whose "wonderful structure and beauty" it alone is in a position to appreciate. Animals and insects on the other hand were determined by particular demarcation, whether in a leopard's sense of smell or a butterfly's dependence upon a single foodplant for its larval phase. Wallace argued time and again, despite his rigorous expositions of Darwin's theory, that to fail to understand this is to open the way for cultural regression. The last chapter of his *Darwinism* (1889) almost forgets the first 14 chapters of uncontroversial exposition as he strives to illuminate the distinctively human quality of the brain: "Because man's physical structure has been developed from an animal form by natural selection, it does not necessarily follow that his mental nature, even though developed *pari passu* with it, has been developed by the same causes only."[38] He goes on to argue that mathematics, music and art "clearly point to the existence in man of something which has not derived from his animal progenitors—something which we may best refer to as being of a spiritual essence or nature, capable of progressive development under favourable conditions."[39] And he draws attention to "the workings within us of a higher nature which has not been developed by means of the struggle for material existence."[40]

Wallace's advocacy of the cultivation of this "higher nature" had led to a serious cooling of relations with Darwin later in their careers. But research a century or so later provides a clue to Wallace's accuracy. For, as Richard Dawkins has argued, natural selection operates at the level of competing genes, not competing organisms. It is genes that are engaged in a struggle for existence, and they therefore do everything they can to increase reproduction in the next generation. This explains altruism; the sacrifice of one individual ensures the survival of other family or tribal members, or as Dawkins seems unable to accept, the species as a whole in the instance of the International Brigades volunteers in the 1930s, or Byron's volunteering to assist in the Greek's War of Independence in the 1820s. Stephen Jay Gould disagreed with Dawkins on this point, and it seems contemptuous of human thought-processes to reduce the species to "gigantic lumbering robots" which harbour gatherings of the great replicator, the gene that is "the basic unit of selfishness." According to Dawkins "we were built as gene machines, created to pass on our genes."[41]

The evolutionist, Loren Eisley, has sharply crystallized the situation of Wallace's reservations on the relation of art to evolution: Darwin "did not, however, supply a valid answer to Wallace's queries. Outside of murmuring about the inherited effects of habit—a contention without scientific validity today—Darwin clung to his original position. Slowly Wallace's challenge was forgotten and a great complacency settled down upon the scientific world."[42] The constant irritation for Darwin was that his younger colleague often anticipated him, and even threatened to surpass in theoretical boldness his own perspectives. It had become clear to Wallace that the superstructural elements in culture play some independent role, even if in his questing spiritualism Wallace never became entirely clear what that may be. Since then many scientists have reduced Darwin's work to a form of nihilism. Indeed when Gould writes triumphantly that "Darwin cut through 2,000 years of philosophy and religion in the most remarkable epigram of the M notebook: 'Plato says in *Phaedo* that 'our imaginary ideas' arise from the preexistence of the soul, are not derivable from experience—read monkeys for preexistence,'" Darwinism has been transformed to absolutist reductionism.[43] Evolutionists before Darwin sensed the dangers, and predominant among these was Robert Chambers.

Chambers, the Edinburgh publisher, in his *Vestiges of natural history of creation* (1844) opened the discussion on evolution, even though Darwin thought it so lightweight as to threaten serious consideration of the issues he was working on. Wallace though was more positive:

> "I well remember the excitement caused by the publication of the *Vestiges*, and the eagerness and delight with which I read it. Although I saw that it really offered no explanation of the process of change of species, yet the view that the change was effected, not though any unimaginable process, but through the known laws and processes of reproduction, commended itself to me as perfectly satisfactory, and as affording the first step towards a more complete and explanatory theory."[44]

Wallace set out to the Amazon specifically to gather facts on this and to test Chambers's hypotheses. George Woodcock argues that "Chambers was interested more in the pattern of evolution than in explaining how it worked. His great appeal was to people who were impatient to solve the riddle of life."[45] However it is more complex than this for Chambers had already intuited the centrality, and mystery, of metamorphosis in nature:

> "Suppose that an ephemeron, hovering over a pool for its one April day of life, were capable of observing the fry of the frog in the water below. In its aged afternoon, having seen no change upon them for such a long time, it would be lit-

tle qualified to conceive that the external branchiae of these creatures were to decay, and be replaced by internal lungs, that feet were to be developed, the tail erased, and the animal then to become a denizen of the land."[46]

H. Lewis McKinney contends Wallace "was quite receptive to the cogent arguments of the heretical *Vestiges*," and he did follow a line of intellectual progress that owes only so much to Darwin.[47] The process of metamorphosis, far more common in insects than other animals, holds the key to this issue. In human culture, this is most clearly traceable in the major works of literature and music where, as I have argued, insect growth takes on a fantastic projection. This has etymological connections connected to the process of transformation—though one is well aware that etymology has declined in academic estimation to such a degree that a recent writer looked out for it as "a common sign that the wheels are coming off the bus!" Gould indeed heaps scorn on philological grubbers in his essay "Glow, Big Glowworm," arguing that it was "Linnaeus himself, father of taxonomy, named the stages of insect development."[48] However in dismissing any interpretations growing from Linnaean etymologies, Gould forgets that the 1858 papers of Darwin and Wallace were presented at the very Linnaean Society itself. That was the cultural context originating the current debates. Remove it from that setting entirely and the subtleties are distorted irreparably.

Chambers hovered at the gates of some new perceptions, but was hindered by his method, limited field work and the state of knowledge in regard to the cosmos and genetics. He wrote:

"Perhaps, with the bulk of men, even those devoted to science, the great difficulty is, after all, in conceiving the particulars of such a process as would be required to advance a fish into a reptile. And yet no difficulty could well be less substantial, seeing that the metamorphosis of the tadpole into the frog—a phenomenon presented to our observation in countless instances every spring—is, in part at least, as thoroughly the transmutation of the fish organization into the reptile, as the supposed change of sauroid fishes into saurian reptiles could ever be."[49]

Here Chambers confused two processes; the relatively linear evolution of animals, from which the frog and clouded tiger salamander are rare deviations, and the transformatory growth of the insect. His projecting the advance of "one grade of animal forms into another" through metamorphosis was wrongheaded. But he sensed there were a number of different processes at work in natural growth—hence his especial interest in the human embryo—and these other processes Darwin did not incorporate into his theoretical method. As Wallace foresaw, in allowing natural selection only a basic function in evolution,

Darwin consigned many aspects of life to the inexplicable, to the realm of a merely illusory independence from nature. The apparent virtues of pluralism and open-endedness came to lack all measure of necessity. It was this that encouraged Wallace to call himself "more Darwinian than Darwin himself" and "the advocate of *pure* Darwinism" from which he felt Darwin had retreated. Paradoxically, such a seemingly deterministic position more freely opens the way for a new resolution of superstructural issues, and for a more advanced dwelling of the human in nature. If one line of human evolution leads back to the great apes, another opens out to the distinctive formative principle of the insect — metamorphosis.

NOTES

1. Wallace (5), 1: 27.
2. Wallace (5), 2: 3.
3. Wallace (5), 2:3.
4. Ruxton, 2135.
5. Marchant, 224–25. Wallace's emphasis.
6. qd. Vorzimmer, 222. In Wallace (18).
7. Wallace (7), 36.
8. Wallace (20), 2:289. I have substituted the modern designations for Wallace's species names.
9. Charles H. Smith, 327.
10. Fichman, 267–68.
11. Wallace (5), 1:401.
12. Wallace (5), 1:401.
13. Wallace (1), 131.
14. Wallace (5), 1:401.
15. Wallace (5), 402.
16. Wallace (14), 716.
17. Wallace (4), 127–30.
18. Dawkins (4), 32.
19. Camerini, 150. Wallace's emphasis.
20. Camerini, 147.
21. George, 251.
22. Camerini, 171–72. Wallace's emphasis.
23. Mallet (4), 1.
24. Darwin (3), 50-51.
25. Nicholson, 1:372. Nicholson's emphasis.
26. Darwin (4), 274.
27. Edey, 70; Bowler (1), 194.
28. Mallet (4), 8.
29. Camerini, 160.

30. qd. Gould (2), 25.

31. Hopkins, 80.

32. Darwin (3), 83–84.

33. Darwin (3), 83.

34. Wallace (5), 1:401.

35. Wallace (4), 429–30.

36. Eisley (1), 306.

37. Eccles, 240. Eccles emphasis.

38. Wallace (2), 463.

39. Wallace (5), 2:474.

40. Wallace (2), 474.

41. Dawkins (4), 19, 36, 199.

42. Eisley (2), 84–85.

43. Gould (2), 25.

44. Wallace (8), 137.

45. Woodcock, 27.

46. Chambers, 207–08.

47. McKinney, 147. Oldroyd argues that Wallace first accepted evolution via *Vestiges*. He certainly was fired by it, but he had read Darwin's *Journal of Researches* in 1842, and re-read it in 1846.

48. Gould (1), 255.

49. Chambers, 212–13. James Burnett, Lord Monboddo, in *Of the Origin and Progress of Language*, had instituted the line of Scottish speculative thought continued by Chambers. Burnett posited the history of the species as the history of the individual writ large.

Chapter Twelve

Of Cells and Mutation

The Scottish visionary, Patrick Geddes, eutopian theorist and town planner, lived for a time in the Edinburgh house once occupied by the prototype for Stevenson's Dr. Jekyll and Mr. Hyde, looking on it so he told his biographer, as an imaginative if ambiguous commitment: "Long ago, I bought this fine old house in High Street and carry on the business. I am a burglar by profession too . . . that's my secret! My diagrams are really skeleton keys, and to ever so many of my colleagues' departments of sciences, philosophies and what not, so I go round even by day and burgle more universities than this one."[1] The method is not dissimilar from that required to identify the intellectual coordinates able to open up the real issue of evolution, that is including the 'subjective' superstructure of the arts and theological structures, and so counteract that mummification diagnosed by Nabokov's Cincinnatus in *Invitation to a Beheading*. This criminal and writer complained that his soul had "grown lazy and accustomed to its snug swaddling clothes," and while he is in this chrysalitic state he fells he is surrounded by "wretched spectres, not people," by Apuleian larvae.[2] Cincinnatus's executioner-to-be is carried off at the end of Nabokov's threaded and almost genetically stranded fable by a woman "like a larva," leaving the prisoner to walk free from his pupal cell to meet with his own kind.[3] The spine of the book is fourfold metamorphic, and constitutes a determining helical paradigm.

One critic has praised the ending of *Invitation* as "Kafka's metamorphosis applied with moral justice," but Roy Pascal has pointed out that there is "no independent narratorial voice" to offer "objective judgment, blame, or apportionment of responsibility.[4] However it is the unerring biological exactitude with which Kafka's beast fables unfurl that carries his writing over from modernism to postmodernism. Nabokov's knowingness at times seems arch and calculating in comparison, for Kafka's transition to postmodernism surpasses not only the epistemological but also the ontological in that, as Brian McHale remarks, ontology is only "a description of *a* universe, not of *the* universe."[5] If the setting for the drama in Kafka's *Metamorphosis* is not an actual cell, then it is cellular in the sense that Gregor Samsa's room has been transformed into an authorial experimental laboratory. As holometabolous insects—those which have a larva unlike their final imago—possess clusters of cells known as imaginal buds, so the writer creates the cell that houses the metaphoric battle for metaphoric freedom, that is perfect realization. Pietro Citati reveals Kafka "again and again, composing with the figures of his unconscious a bestiary just as immense as a medieval one," and the fact that he wrote this fable not long before the First War while Nabokov wrote his *Invitation* some four years before the Second, indicates the evolutionary pressures on the historical scale, the wars in one analysis being the outgrowth of what Geddes described as nations "dazzled by that 'impressive nature-myth' of tooth-and-claw competition."[6] This is the brutal context in which Nietzsche's prescription in *Twilight of the Idols* is fought out: "How the 'true world' finally became a fable."[7] So the writers' cell is suspended over the void; it is the organism out of which new life-metaphors strive to emerge. Nabokov characterizes *Invitation* as "a violin in a void"; Kafka enacts the way "vermin is born of the void," both recalling Matthew Arnold's cynical view of Shelley as that "beautiful and ineffectual angel, beating in the void his luminous wings in vain."[8]

It can be argued that Kafka's bug is no known creature, deliberately and decisively fictitious. But while he opposed its pictorial representation, and its interior processes of change are the lynch-pins of the plot, its nature does add some crucial nuances. Nabokov in his *Lectures on Literature* settled for "merely a big beetle."[9] The author himself refers to "the black beetle of my story" even though it is in fact brown.[10] Although Willa and Edwin Muir's pioneering and standard translation of 'Ungeziefer' as "dung-beetle" has been criticized, nonetheless the fact that Samsa ends is urban days among ash- and garbage-cans suggest this is not far wide of the mark. Moreover since the charwoman represents the brashly realistic world in the latter part of the story and is permitted to call him "Mistkäfer," dung-beetle, he might be classified as a member of that yellowish-brown chafer family—not least because he does after all chafe at everything. The "slight itching up on his belly" he feels while contemplating his promotion from warehouse hand—only claw-like terminal appendages now!—to salesman duplicated Kafka's extreme touchiness when "my strength no longer suffices for another sentence."[11] And there is perhaps an intimation of the phrase "to start again from scratch" in Gregor's collapse into a state without beginning nor end, no sustenance save scaping the floor. Nor is there prospect of escape even in the route sounded out by the most naked intervention of the narrator, ironical to the point of jeering—"he couldn't have flown away, could he?"—which at that early stage of transformation was indeed impossible though later, as Nabokov points out, he might have become aware of his sharded wings.[12]

The almost cruel forcefulness of the narratorial intervention is explicable only in terms of Gregor's being in a chrysalitic situation in Chapter 2, as I will explain in a moment. His injuries will make his wings useless for he is destroyed by his pupal circumstances—family and position—before he reaches that imago point. Instead his coleopterous physiology gives him a certain tank-like quality. He, like Kafka himself, may be awaiting that liberating experience enjoyed by Cincinnatus when he wearies of the grounded book he is reading, *Quercus*, whose author sits enthroned "somewhere among the topmost branches" of his parnassian and safely stolid tree. The execution slab is also of oak, but at the conclusion as he flees his doom, trees are falling all around him. An inkling of rebirth symbolically arrives in the shape of "a large dummy acorn, twice as large as life" which bounces on his prison blanket "fitting its cork cup as snugly as and egg."[13]

Sokel distinguishes twin paradoxes in Gregor's condition. His human feelings remain even while he cannot communicate them; and his daunting exterior conceals a new inner vulnerability.[14] One might add that the beetle form expresses this perfectly since the hard wing-cases, elytra, cover the membranous second pair beneath them. There is a third contradiction in that while as a human he had no aesthetic sensibility, as an insect he is drawn to music, associated as it is with his attraction to his sister. But the real battles in Kafka take place not within the sphere of paradox, dualism or even dialectic, but between the quaternal and triple. Roy Pascal intimates the absoluteness of the conflict when he foresees the demise of the traditional fable whose function "was to allow us to understand life, to order and label its manifestations, to teach us practical wisdom that will serve to guide our behaviour," whereas "this fable of Kafka's does not illuminate the mind but terrifies and confuses."[15] However Nabokov discovers a paradigmatic cluster of triplets:

> "The story is divided into three parts. There are three doors to Gregor's room. His family consists of three people. Three servants appear in the course of the story. Three lodgers have three beards. Three Samsas write three letters."[16]

The numerical tension arises, though, from there being four phases in his transmutation (egg, larva, chrysalis, beetle) while only three of these are properly completed. He can never enjoy his imaginal situation.

Contrary to Nabokov's overly generalized entomological comments, Gregor is probably still in grub form at the start having emerged from his egg of convention to find he has "numerous legs" waving in the air. A beetle has but six legs where its larva has myriad. Exactly halfway through the fable, he is recounting how he "especially enjoyed hanging suspended from the ceiling," and this clearly represents the final stage of a pupation.[17] It is the second of three chapters that describes this process, and it is here he tries out the antennae which in Coleoptera, as in most Lepidoptera, are visibly extant at the chrysalis moment. However the doom of the threefold haunts Samsa. As with all insects, the beetle's body is segmented into three sections, and his fate is to follow this configuration. His last memory is of the clock striking three. He leaves the room only three times, but this would appear to seal his destiny for each of his purposive acts has a fourfold aspect. "It cost him four hours' labour" to move a sheet to the sofa to hide from his sister, and when the furniture is being moved he changes direction four times to conceal himself.[18] Groups of four words determine the objective situation. The clerk pronounces Samsa's speech "the sub-human gibbering of a beast—'das war eine Tierstimme.'" As F.D. Luke concludes, the clerk "here utters in four words Kafka's whole criticism both of himself and mankind."[19] Moreover Samsa's realistic summary of his dilemma is quadruple—"Es war kein Traum"—and his pathetic later question-cum explanation emerges from quaternity: "War er ein Tier, da ihn Musik so ergriff?"[20] His tetrapteran nature is ultimately futile, for the fable ends in a blazing litany of the triple in which he immolates himself and out of which the daughter arises from her nymphal larva, though with a shade too much of merely animal high spirits. Two lines of character development have been at diagonal odds in the tale. With Gregor's mutation, it has been that of the holometabolous insect; there is no apparent connection between his final beast form and his human existence. His musical sister, on the other hand, has undergone a threefold hemimetabolous change, from egg to larva and thence by gradations to imago. In entomological terms the nymph or naiad requires no specific period of transformatory preparation, no pupa, but evolves to full maturity by way of a series of ecdyses, or larval moults.

If the death of the fable genre is to be surmised—and it is an expiry that will rather be an eternal return to Apuleius—then the work threatening its demise is *A Report to an Academy*. Here narrator and protagonist are one, since Kafka gives the floor to the ape. As the animal says, within his five year education "I managed to reach the cultural level of an average European. In itself that might be nothing to speak of . . ."[21] But this is not just a formal resolution of narrative technique, for in calling the artistic bluff by putting humanity's ancestral animal at the very centre, Kafka has raised the fable to the point of evolutionary enigma. There is a paradox at the end which in the light of subsequent scientific knowledge confirms the writer's ironical intuition. The ape is reported as having "a half-trained chimpanzee" as a pet, albeit one he cannot bear to look at in the coldness of daylight so pitiful is its broken spirit.[22] There is a double and ultimate burlesque in that the simian species nearest the human in its DNA constitution is the chimpanzee. In the context of the Great War during which this fable was written, the ape can claim not only historical primacy over the human, but can report that it has surpassed its evolutionary offspring in culture and morality.

One critic has defined the difference between Nabokov and Kafka as lying in the fact that Nabokov "possessed a fierce, sometimes hysterical, belief in the ultimate triumph of the individual artist (the two words were inseparable for Nabokov) over the absurdity of perceived existence."[23] Edwin Muir quite lacked the dandyish individualism of the Russo-American author and could have had scant sympathy with his deracinated vision, yet he had much in common with this belief. When Hermann Broch translated Muir's poem originally entitled "Transmutation" and which finally became "The Threefold Place," he rendered its title as "Verwandlung," identical to that of Kafka's tale.[24] Muir undertook something akin to what he proposed in his poem "To Franz Kafka" to be "Eternity's secret script, the saving proof," though not so much in metamorphosis as in the Transfiguration.[25] He intimated an heroic age of deeper relationship between human and beast echoed in the original title of his autobiography, *The Story and the Fable*, and glimpsed in the childhood fusion of "the ordinary and the fabulous" in Scandinavian Orkney. He located the fable of each person's individuality within life's "unexpected and yet incontestable meaning which runs in the teeth of ordinary experience, perfectly coherent, yet depending on a different system of connected relations from that by which we consciously live."[26] This is an extremely democratic and almost unmediated fable, one not so far as may appear from the postmodern prescriptive "a change of paradigm always and necessarily involves a change of world."[27]

Indeed Muir's characterization of modernist theory is still potent: "we want a little chaos in our order, we are no longer lugubrious and apocalyptic over evidence which shakes a Design; on the contrary, we are pleased, as if in some way a little part of a wearisome burden had been lifted from our spirit."[28] There is relief, liberation, in randomness, not least in quantum mechanics. H.J. Blackham has proposed that "the longer complex fables go beyond the simple images of the Aesopic fable, but simplify the abstractions they represent in a kind of model, as modern scientists build models of their invisible conceptual entities, based on measurements."[29] As we have seen, this is the case in modern fables where number symbolism plays a major part. Nabokov's "The Aurelian" is another of his city fables. It begins with a torrent of numbers; the street on which the aurelian has his shop is determined by "luring aside one of the trolley-car numbers" and has a small square with four benches; there are four stores on the right hand side culminating in "all of a sudden, a butterfly store." There are three more "ordinary shops" before a fourth, a bar where four people play for stakes of four drinks.[30] Pilgram is the lepidopterist whose progress the fable charts. But the four sections of the tale end not in metamorphic completeness for him, but in death from nervous palp-itations brought on by his over-eager anticipation of an illicit butterfly-hunting expedition. Nabokov viewed his writing as "a kind of merging be-tween two things, between the precision of poetry and the excitement of pure science."[31] His answer to Darwin's perceived reductionism lies in the excess and luxury of nature as he interprets it. So the markings on a butterfly wings in imitation of a leaf or of a dewdrop go beyond mere need in their detail. Art and literature arise from such superfluity and abundance, but they also con-tain an almost genetic message connecting us to the insect.

NOTES

1. Defries, 241. Geddes always distinguished his eutopia, 'good place,' from William Morris's utopia, 'no place.'

2. Nabokov (1), 30–32. For the Apuleian larvae, see Chapter 3.

3. Nabokov (1), 191.

4. Field, 146; Pascal, 57.

5. McHale, 27. The emphases are McHale's.

6. Citati, 59; Boardman, 270.

7. Rosen, 200.

8. Nabokov (1), 9; Kafka (2), 99; Arnold, 237. There is a curious pre-echo of Kafka'a inchoate characters without start or finish in the noted phrase of the Edin-burgh geologist, James Hutton: "no vestige of a beginning—no prospect of an end (Hutton, 304)."

A passage from one of Shelley's letters evokes a more playful concept of the trans-formed void: "I live here like an insect that sports in a transient sunbeam, which the next cloud shall obscure for ever . . . Burns says, you know, "Pleasures are like pop-pies spread,/You seize the flower—the bloom is fled;/Or like the snow-falls in the river,/A moment white—then lost for ever (Shelley, 383)." The Burns contains inac-curacies.

9. Nabokov (2), 260.

10. Brod, 304.

11. Kafka (1), 90; Brod, 39.

12. Kafka (1), 107; Nabokov (2), 39.

13. Nabokov (1), 105, 107.

14. Sokel, 210–11, 212–13.

15. Pascal, 150.

16. Nabokov (2), 282.

17. Kafka (1), 89, 115.
18. Kafka (1), 114, 118.
19. Luke, 242.
20. Kafka (3), 130.
21. Kafka (1), 258.
22. Kafka (1), 259.
23. Field, 146.
24. For a full exploration of the translation, see Gaskill.
25. Muir (1), 216.
26. Muir (4), 12, 49.
27. Battersby, 120.
28. Muir (3), 250.
29. Blackham, xvi.
30. Nabokov (3), 75.
31. Nabokov (5), 10.

Chapter Thirteen

Ancient and Medieval Cosmology

Medieval cosmology, not unlike Buddhism and Hinduism, lay great store by manifestations of the fourfold. Many structures dependent upon this number had their origins in the pre-Socratics. Empedocles lighted on the four elements in the course of a Parmenidian dialectic of Love and Hate. But Pythagoras had already conceived the more creative, or at least transcendent, 'music of the spheres' which was to become central, of course, to medieval cosmology. The earliest number symbolism occurred in Babylon, and the *zikkyrats* varied in numbers of stages between 4 and 7. The Babylonian system had 4 directions, lunar phases, winds, seasons, watches of the day and night, 4 elements, humours and cardinal virtues. Almost all peoples of antiquity possessed a name for the Deity composed of four letters. Among Gnostics, the Supreme Being was denoted by the number 4, and in the *Acts of Thomas* when passing through the fourth gate the initiate is referred to as a "body released by a pentad." The Hebrew *Tetragrammaton*, or 4-lettered mystery, is the name of the Creative Power. It includes the past, present and future forms of the verb 'to be', and was revered as a symbol of the immutable I AM. Jehovah is INVH. With regard to Christian innovations in number science, central

"was the identification of the spiritual-temporal duality with the archetypal numbers 3 and 4. Four, by the known analogues of the 4 winds, the 4 elements, the 4 seasons, and the 4 rivers, is specifically the number of the mundane sphere; and as the first 3 days of creation foreshadow the Trinity, so the fourth is the 'type of man'. Mystically, the fact that man is a tetrad is evidenced in the name, *Adam*, whose letters are the 4 winds. For this reason, knowledge of the divine things is disseminated throughout the world by the 4 gospels, evangelists or beasts, emblemized by the 4 extremities of the cross, the 4-fold division of

Christ's clothing, and the 4 virtues, or forms of love, as Augustine names them."[1]

The centrality of the number 4 is not as mystical as may appear at first sight. Apart from 2, all prime numbers are odd. There are those that are 1 less than a multiple of 4—3, 7, 11 and 19; and those that are 1 more than multiple of 4—5, 13, 17. The Pythagoreans, said Nicomachus, call the number 4 "the greatest miracle," "a God after another manner," "a manifold divinity," the "fountain of Nature" and its "key bearer."[2] The first globe, made by the Greek Crates in 150 BC showed 4 Continents separated by oceans. This was not because the existence of N. and S. America or Australia was even guessed, but because it was thought that 4 continents were necessary for the sake of balance. Significantly Pythagoras, heir to the science of mathematics originating in Babylonia 3,000 years earlier, hypothesized this as a corollary to the invention of Mathematics and as laying the basis for the discovery of the octave: "The Pythagoreans first stated the mathematical ratios involved in musical harmonic relationships, and these all turned out to be contained within the number 10, and to be expressible as ratios of the first four numbers."[3] For the Greeks, number was the thing itself, and not merely an attribute of the object. Indeed, to them, "the tetrad was the root principle of all things because it was the number of the elements."[4] This was emphasized by Schopenhauer who wrote that Pythagoras "implanted in our mind the quaternary number, the source and root of eternally flowing creation."[5] Plato's *Timaeus*, which combines mathematics, music and mysticism, demonstrates the world is a copy of a perfect model comprising four primary bodies created by the Demiurge who fixed the world in definite quantities and properties. The fusion is of "the world soul (a religious concept), the regulation of the cosmos (a concept of physics), world harmony (a musical concept) and the soul of man (a psychological concept)"[6] This is the source of concepts of Creation and the basis for a scientific cosmology.

There is much in the circulation of this fourfold number structure throughout the Graeco-Roman period into the Middle Ages that resembles DNA as a replicator, making copies of itself. As Dawkins remarked: "If individuals live in a social climate in which imitation is common, this corresponds to a cellular climate rich in enzymes for copying DNA."[7] So this pattern of thinking spread throughout medieval society, which in its conformity is characterized by a high degree of imitative behaviour. Some of this was clearly inherited from earlier systems of religion, and in a meme sense was transmitted into France where in the seventeenth century drama Corneille's "conception of the nature of man is defined with the mathematical precision of Descartes."[8]

Number is the word made flesh in the medieval world, or alternatively it is the very skeleton of the world order. It has not yet become the merely common currency it is reduced to when universal trade is the norm. And numerical sacredness is enshrined in the Judeo-Christian Trinity. As Edmund Reiss remarks: "Such theoretical developments as the formulation of the concept of Trinity by Clement of Alexandria and his pupils, Origen and Hippolytus, with its mystical sense of relationship of three in one, allowed for a mystical sense of number in orthodox Christian doctrine."[9] This, though, had an ancient predecessor in the Gilgamesh epic from sometime before 2000 BCE, which portrayed a threefold world of Anu (the heavens), Enlil (the earth) and Ea (the waters). This is the most ancient triad, and in the course of time a second was added: Sin (moon), Shamash (sun) and Adad (storm and water). A fourth region was occupied by God. Four and three together constitute the Hebrew major sacred number—seven, which embodies certainty. The number occurs 54 times in Revelation. Frequently there are ratios of 4:3 and 3:4. So in the seven epistles the first three form one group, the last four another. In the seven seals, the peculiarity of the revelation of the seventh marks the change to a new group. However Isbon Beckwith argues "the theory that seven originally owes its sacredness to the fact that it is the sum of three and four is in itself improbable; at any rate there is not indication of this thought with our author; nowhere so far as the numbers are expressly named does he bring seven into connection with three and four, as constituent parts."[10] Jung has much that is illuminating to say on this:

"I cannot omit calling attention to the interesting fact that whereas the central Christian symbolism is a Trinity, the formula of the unconscious mind is a quaternity. . . . It could easily be inferred that the fourth [aspect] represents the devil. . . . From an orthodox standpoint, therefore, the natural quaternity could be declared to be 'diabolica fraus' and the capital piece of evidence would be the assimilation of the fourth aspect which represents the reprehensible part of the Christian cosmos."[11]

The Manichees were tetradites, believing the godhead to consist of 4 persons and Satan coterminous with God. However in the end, Jung fails to do more in his work as a whole than throw out vague hints. He remarks that while there is not an exact, one to one, literal basis for the journey of the soul, there is some unknown analogy, a metaphor. But he was unable to clarify what that metaphor was. While he opposed Freud's relating everything back to the "instinctual processes conditioned by the body," proposing instead the "sovereignty of the psyche," he has to retreat to conclusions such as "the—at present—insoluble task of reducing everything psychic to something definitely physical."[12]

Medieval thought found a resemblance between the 4 cardinal virtues (prudence, justice, fortitude and temperance) and the core of education syllabus, the Quadrivium. At the same time the Trivium was shadowed by the theological virtues (virtue, faith, and hope with charity). Almost despite the Trinity, the Bible is rich in the number four emphasizing the syncretistic view that there is a relation between number structures and God. When the Jews were dispersed on losing their political independence, they maintained their unity through the formalizations of the books of Ezekiel and Daniel, together with their New Testament equivalent the Book of Revelation. There are Ezekiel's 4 winged creatures which Byrhtferth drew on and which can be taken as a prophecy of the four evangelists, together with the chariot's four wheels within wheels symbolizing the containment of the New Testament within the Old, the 4 winds striving on the sea in Daniel, the fourfold compensation in Exodus, 2 Samuel and Luke, together with the fact of there being 4 gospels. Maren-Sofia Röstvig observes:

> "It is fairly well known that the 4 rivers of Paradise were considered as a type or prophecy of the gospels, but it is nevertheless surprising that the same prophecy could be attributed to a purely literary structure simply because of the presence in this structure of the number 4. Lamentations consists of 5 chapters, all but the last of which display an alphabetical technique of composition. Because the contents were universally interpreted as a prophecy of the passion of Christ, the structural numbers (22 and 4) were interpreted accordingly. The form was taken to indicate that Jeremiah's lamentation was for the sins of the whole world, as the world is governed by the number 4 (the 4 seasons, the 4 humours, etc.)."[13]

The Book of Revelation is shot through with the significance of numbers: 3, 4, 7 and 12. Here is enunciated Last Things, and the 7 unsealings begin with the judgments of Four Horsemen of the Apocalypse. Even the structure of Revelation falls into quaternal shape: the messages (i-iii), the unsealings (iv-xi), the trumpets (viii–xiv) and the bowls (xv-xxii). There are four forms of evil: false apostles, the false synagogue, the false prophet and godless king, and the bowls. The four living creatures (iv, 6) are traditional, while the four cherubim bear Jehovah's throne-car into the four quarters of the earth. And the Holy City is solidly foursquare. This all runs dialectically with the number three, which is one of the scriptual numbers signifying adequateness. So in Revelation, there are the 3 plagues (ix, 18), the 3 woes (viii, 13) the effects of the earthquake (xvi, 19), and the 3 gates on each side of the wall of the New Jerusalem (xxi, 13) which derive from Ezekiel.

The 1st century CE Theon of Smyrna has a chapter "On the *tetraktus* and the decad" in *Expositio rerum mathematicarum ad legendum Platonem utilium* where he relates the numbers 1–4 to the Magnitudes—point, line, suface, solid, and to what he calls Simple bodies with their figures—fire (pyramid), air (octahedron), water (icosahedron) and earth (cube). He then extends this quaternity; to Living things, being seed, growth in length, in breadth, in thickness; to Societies, man, village, city, nation; to Faculties, reason, knowledge, opinion, sensation; to the four seasons; to the four ages of humanity and to the parts of the human being which he conceives of as the body and three parts of the soul. These were all attempts to interrelate human development with psychology and the environment, the microcosm with the macrocosm. They are repeated in differing forms over a millennium. But Ovid's *Metamorphoses*, written during the time of Christ's life, had pre-dated the Theon, and surpassing many formulaic presentations, caught the transformations of the fourfold processes:

> "don't you see the year passing through a succession of four seasons, thus imitating our own life? In the early spring, it is tender and full of sap, like the age of childhood. Then the crops, in shining trim but still delicate, shoot up in the fields and, though they are not yet stout and strong, fill the farmers with joyous hopes. Everything is in flower, the fertile earth gay with brightly-coloured blossoms, but there is as yet no sturdiness in the leaves. Spring past, the year grows more robust and, moving on into summer, becomes like a strong young man. There is no time hardier than this, none richer, none so hot and fiery. Autumn takes over when the ardour of youth is gone, a season ripe and mellow, in temper midway between youth and age, with a sprinkling of grey hairs at its temples. Then aged winter comes shivering in, with tottering steps, its hair all gone, or what it has turned white...
>
> The everlasting universe contains four elements that give rise to bodies. . . . Though these four elements are distinct from each other in space, yet they are all derived from one another, and are resolved back into themselves. Earth is broken up and refined into liquid water, water becoming still less substantial changes into air and wind, and air too, being already of the finest texture, flashes upwards when it loses weight, into the fiery atmosphere above. Then the process is reversed, and the elements are restored again in the same order: fire condenses and thickens into air, air into water, and water, under pressure, produces earth."[14]

The Venerable Bede in 725 expounded what came to be known as the Physical and Physiological Fours in his *De Temporum Ratione*. This contends that "human life is fourfold, governed in turn by each of the four humours in the body, and harmonized with the macrocosmic order of the seasons and the elements through the same system of qualities that governs them all)."[15]

In a late flowering of the Anglo-Saxon culture, Byrhtferth's *Manual* of 1011 projected the ancient elements into what one commentator has called "a blue-print of universal nature."[16] It is quadrifoliate to the point of obsession, but has a similar majesty to the brackets, stringcourses, finials and spandrels of the great medieval cathedrals whose design though was Norman. It is this type of structured thought that has gone into their conception, even if it is too schematic and assured for the modern age with its quantum vision. In Byrhtferth, the four elements and temperaments underpin the whole, together with the four ages of humanity:

"boyhood	early manhood	manhood	old age
spring	summer	autumn	winter
moist & hot	hot & dry	dry & cold	cold & moist
air & blood	fire & red bile	earth & melancholy	water & phlegm"[17]

The four elements are in turn related to the four virtues—justice, prudence, temperance and fortitude. In a fragmented sentence on the putative significance of his book, Byrhtferth writes that "the mysteries of the allegorical meaning of this slight work, we whose powers are limited, must with God working through us perform now with ampler result the task we have undertaken with hurried pen."[18]

He goes on to distinguish 4 scansions or caesuras in poetry. Later, it was the Platonists who kept this schema alive as the astrologers introduced the 7 ages of man in the 12th and 13th centuries. Anthony Nixon, though, was connecting the 4 ages of man to the 4 seasons as late as 1612 in his *The Dignitie of Man*.

Beowulf has lines of 4 stresses divided into two half lines with two stresses in each. Also the greatest Eddaic poem, the *Volospá*, tends to 4 syllables. There is an interesting technical distinction between Icelandic and Anglo-Saxon poetry in this regard. In the Icelandic, there is an exact relation between sense and poetry with sentences coinciding with the length of lines, whereas in Anglo-Saxon the sentences burst beyond the lines and even the verses. It is worth remarking that the German Romantic lyric in the style of the artificial folk-song usually follows the 4-line stanza of the original folk-song, and there is the 4-beat trochee of the Spanish drama and romances, and the 4-beat rhyming verse of the *Roman de la Rose*. Byrhtferth, following through his various *quadrans* which by now have taken on the character of the tetrapterous, elucidates the four criteria for deciding on Easter, before breaking into a rhapsody on the number 4, in the course of a general disquisition on number:

"The number four is a perfect number, and it is adorned with four virtues—righteousness, temperance, fortitude and prudence. The number is also crowned with the four seasons of the year whose names are: spring, summer, autumn and winter. It is also adorned with the doctrines of the four Evangelists, who are said to be the four animals in the book of Ezechiel, the famous prophet. The number four is reverently upheld by the four letters in the name of Christ, that is to say,

D.E.U.S., and likewise by the name of the first created man, namely, Adam. Fittingly it has an attraction which I do not think ought to be passed over in silence—I mean [the fact that there are] two equinoxes and two solstices. There are indeed four principal winds, whose names are these: the east, west, north and south, as the Psalmist sings: *A solis ortu, et occasu; ab aquilone, et mari* (Psalm cvi. 3). If these parts are carefully studied, they will be found in the name of Adam according to Greek numeration."[19]

During his discourse on Four, Byrhtferth moves between Anglo-Saxon and Latin, finally staying in Latin to the end of the book as the language best suited to abstract thought. He reverts to the vernacular only for clarifications and elaboration.

In *The Kingis Quair* of *James Stewart*, "the author reviews his life only in order to illustrate the single significant pattern that he has discovered in it."[20] The poet's lady appears in Stanza 40, the number having Biblical implications of constraint and suffering. (James I of Scotland was imprisoned by the English for 24 years, though his eventual wife for whom this poem was written, was English). The bodily climacteric arrives at Stanza 49, seven being the body's number, and 49 also being the square of the sum of the male and female numbers. The final stanza is number 196, 4 × the orgasmic 49. As Gregory Kratzmann puts it, "the poem is presented as the outcome of the poet's review of his life ('and all myn auenture/I gan ouerhayle') and its action creates very effectively the illusion that he is reliving his experience in imagination."[21] Number symbolism is to found in most fourteenth and fifteenth century texts, outside of Chaucer. Chaucer is one of the first *literary* writers, just as later Ben Jonson was to be the first self-conscious author as such. John MacQueen points to Chaucer's numbers as being "decorative rather than functional: the main purpose seems to be no more than the preservation of symmetry in the narrative."[22] A similar situation holds with Edmund Spenser whose *Shepheardes Calendar* cycle begins in January and the birth of Christ:

"The tempering effected by/the 4 of the seasons, the 4 of the passions and the humours, and the 4 of the gospels would have presented itself as a theme obviously connected with such a calendar, and the claim of the envoy that the calendar comprises all of time would seem reasonable to all who remembered St. Augustine's alignment of the 4 ages of man and the 4 seasons with the ages in the universal history of man."[23]

Fowler has shown how Book IV of his *The Faerie Queene* "is saturated with numerology, as if Spenser had been possessed with a vision of number as a chain of concord binding the world's workmanship together," and the narrative "is throughout dependent upon the tetrad, concerned as a symbol of friendship and concord."[24] Alliances between evil characters engender only caricatures of the fourfold. And further, Fowler points out that "the guardian deity of the fourth book is Mercury, the fourth god in the planetary week (the guardian of Wednesday, or *dies Mercurii*); and Mercury is, above all, the god of concord, and of that reason by which the whole world is ordered. . . . Four was his number, and so he became known as the *quadratus deus* and the god of true proportion."[25]

John MacQueen maintains that "the most interesting and elaborate British examples of literary numerology of the period come from areas well to the north and west of the metropolitan culture within which he flourished."[26] So *Sir Gawayne and the Grene Knight* unexceptionally has four parts echoing the 4 seasons of the year. The seasons have to pass before Gawayne has to go to the Green Chapel for his trial by the Green Knight representing the old order. This Lord of the Castle goes hunting for 3 days before Gawayne's trial which occurs on the fourth day. The whole poem is 'sprung' by the flexibile alliterative measure of the fourteenth century. This has 4 heavily stressed syllables in the line with 3 of them alliterating, two on each side of the pause. These stressed syllables can, as Francis Berry noted

> "be very freely moved about in relation to each other inside the line on account of the large freedom of choice as to the number of unstressed syllables intervening between them. This allows a line with four 'blows' of variable, often very heavy, weight (occasionally two of them immediately succeed each other to make a centre of force), and this offered the writer a variety of rhythmic contour, as well as a varied length of line."[27]

Likewise there was widespread use of quatrains in medieval French poetry, along with a fourfold structure of the poetry. So Chrétien de Troyes builds his *Erec* and *Yvain* around four distinct sections. The narrative begins at King Arthur's court and introduces the hero as he wins his bride, this being followed by the description of festivities; a crisis develops between husband and wife; after a series of adventures hero and heroine are reconciled; and there is a final elaborate interlude with fantastic and supernatural events.

Dante's poetry, as has often been remarked, is closely woven around the numerical. His *La Vita Nuova* is locates the prospect of a new life through love which is identified with his perfect number, read through the Christian and Arabic tradition. Number also defines individual identity. In the first few lines, the female object of sacred devotion is characterized as 3×3, the cube of the Trinity. The poet first saw Beatrice on the ninth hour of his ninth year:

> "Nine times already since my birth had the bright heavens almost returned to the self-same point in its own revolution, when the glorious lady of my soul first ap-

peared before my eyes. She was called Beatrice by many, who did not understand the significance of her name. She had been alive for that span during which the starry heaven had moved towards the East on twelfth of a degree, so that she appeared to me at about the beginning of her ninth year. And I beheld her almost at the end of my ninth year."[28]

Her death occurs at 3^3, perfectly within the limits of "this number which was her constant friend":

"I say that according to the Arabic computation, her most noble soul departed at the first hour of the ninth day of the month; and by the usage of Syria, she departed in the ninth month of the year because the first month there is Tixryn, which is our October. And according to our system, she died in that year of our reckoning, the year of Our Lord, in which the perfect number had been counted nine times in that century in which she was set in our world. And this was the thirteenth century of the Christians.

And perhaps this is why this number was her constant friend; namely that since according to Ptolemy and Christian truth there are nine heavens which move and, by the general opinion of astrologers, these heavens exert an influence upon the lower world through their conjunctions, this number was associated with her, to signify that at her birth all nine moveable heavens were in perfect conjunction.

This is one way of reasoning. But considering the matter more deeply and by way of absolute Truth, this numeral was her very self, her essence—I speak by way of metaphor as I will explain. The number three is the root of nine, for independently when multiplied by itself it makes nine and by rote we know that 3 $\times$ 3 makes 9. If then the 3 by itself is the only factor of 9, and the sole factor of miracles is the Three in the Father, Son and Holy Ghost who are the Three-in-One, then was my lady attended by the number 9 to the intent that it might be understood she was a Nine, that is a miracle, the root of which is none other than the miraculous Trinity. Perhaps some more dialectical mind might discern a still more subtle reason for it; but this is what I see that pleases me most."[29]

Metaphor in Dante then is not, as in Shakespeare, embodied in imagery, but in the higher abstraction of number. The symmetry of Shakespeare's plays is revealed through an imperious mind, whereas Dantean structure is embodied in his sacred lady.

However the relationship between the writer, his matter, and the reader involves yet another numerical ratio which derives from the Augustinian axiom that the theme of Scripture is charity, and that the purpose of all Christian interpretation of texts is to disclose and promulgate this theme. So the *Divina Commedia* was usually read four ways, and the embedded strata can be interpreted: literally (the ovum, in my terms), morally (the larva), mystically (the chrysalis) and anagogically (the psyche). Then leaving aside the first canto which is by way of a preface to the whole poem, each of the 3 parts of the *Divine Comedy* holds 33 cantos, and it is a critical commonplace that *terza rima* is a homage to the Trinity. George Steiner has located the apotheosis of triplicity at its conclusion: "But in an ultimate *ricorso*, as we shall see, the final canto of the *Paradiso* contains three explicit allusions to Virgil, where this threefold recall takes its august place within the trinitarian algorithm, within the sacred triplicities of the formal-theological design. (There is a trinitarian 'phonetic' in the very name of Bea*trice*)."[30] Virgil's *Aeneid* itself is divided into three groups of four books. Each of the twelve books have three divisions which are themselves usually broken into three parts. As George Duckworth notes: "The *Aeneid* is thus a trilogy with the first four books, the tragedy of Dido, and the last four books, the tragedy of Turnus, enclosing in a framework pattern the central portion, where living Homeric episodes (games, trip to the underworld, catalogue, description of a shield) are re-worked and transformed for the glorification of Rome and its history, the portrayal of ancient Italy, and the praise of Augustus and the new Golden Age."[31] More generally, Dante perceives a threefold pattern to human life. He separates out a life-span of 70 years into a period up to the age of 25 (*adolescenza*), a phase of maturity (*gioventute*), and an age of decline from 45–70 (*senettute*). This echoes the biological stages of growth defined by Aristotle in *De Anima*, for whom an underlying pattern in nature provides a pattern of growth, development and reproduction. These arise from the vegetative section of the soul. Two other parts of the Aristotelian soul are the sensitive and intellectual. This threefold scheme applies to humans, animals and plants.

Schelling in his *On Dante in Relation to Philosophy* argues that "just as for recent drama the five act form is regarded as usual, because each event can be seen in its beginnings, continuation and culmination, progress to completion and actual ending; so for the higher prophetic poetry which expresses a whole age the trichotomy of Dante's is conceivable as a general form, but one whose filling out would be endlessly varied as it is revitalized by the power of original invention."[32] It is probable that Dantean triplicity is associated with the cult of the Virgin for the triform is traditionally emblematic of woman. W.K.C. Guthrie remarks on "some curious notions about the primacy of the number three which clearly antedate the beginnings of philosophical thought."[33] So both the three Fates and the three Hesperides are the Triple-Moon goddess in her death aspect. Mythically, Cronus castrated Uranus at the behest of Mother Earth who provided him with a flint sickle:

"But drops of blood flowing from the wound fell upon Mother Earth, and she bore the Three Erinnyes, furies who avenge crimes of parricide and perjury—by name Alecto, Tisiphone, and Megaera. The nymphs of the ash-tree, called the

Meliae, also sprang from that blood. . . . The ash-nymphs are the Three Furies in more gracious mood: the sacred king was dedicate to the ash-tree, originally used in rain-making ceremonies. In Scandinavia it became the tree of universal magic; the three Norns, or Fates, dispensed justice under an ash which Odin, on claiming the fatherhood of mankind, made his magical steed."[34]

Among a multitude of other threefold females are *Macbeth's* three witches, the three symbolic daughters of Thibaut d'Arc in Schiller's *The Maid of Orleans*, Lear's three daughters, the triple naiads in Wordsworth's *Triad*, the three Horai, the trio of emanations of Gorgon and Circe's three sirens, and the Scandinavian three Norns (Fates) who attend to the root of Asgard the abode of the gods.

It is most likely that the threefold was the predominant organising number of early matriarchal societies, which was then carried over into Judeo-Christianity. D.H. Lawrence has a rhapsodic meditation of the number in *Apocalypse*, his final book and the most concentrated assault on contemporary materialism and anti-spiritual attitudes to this day:

"Three was the sacred number: it is still, for it is the number of the Trinity: it is the number of the nature of God. . . .
   Again, we know that Anaximander, in the fifth century, conceived of the Boundless, the infinite substance, as having its two 'elements', the hot and the cold, the dry and the moist, or fire and the dark, the great 'pair', on either side of it, in the first primordial creation. These three were the beginnings of things. This idea lies at the back of the very ancient division of the *living* cosmos into three, before the idea of God was separated out. . . . It is paralleled in Genesis, where we have a God creation, in the division into heaven, and earth, and water: the first three *created* elements, presupposing a God who creates. The ancient threefold division of the living heavens, the Chaldean, is made when the heavens themselves are divine, and not merely God-inhabited. Before men felt any need of Gods or gods, while the vast heavens lived of themselves and lived breast to breast with man, the Chaldeans gazed up in the religious rapture. And then, by some strange intuition, they divided the heavens into three sections. And then they really *knew* the stars, as the stars have never been known since.
   Later when a God or Maker or ruler of the skies was invented or discovered, then the heavens were divided into the four quarters, the old four quarters that lasted so long. . . .
   Three is the number of things divine, and four is the number of creation."[35]

The triple historically gave way to the quaternal, because the evolution of imagination and invention "depends upon the fourfold activity." Yet, as Lawrence argues, the science of mathematics awaits its perfection through union with human life. From a very different and theological angle, the Scottish Borderer Thomas Boston, in a work which had a wide readership among ordinary eighteenth century folk, intimated a fourfold key to human psychology in his *Human Nature in its Fourfold State of Primitive Integrity, Entire Deprivation, Begun Recovery, and Consummate Happiness or Misery*. The absorption of the tetradic structure into an idiosyncratic plan for universal regeneration (or degeneration into misery!) is an ingenious intellectual achievement from the Robinson Crusoe/ Alexander Selkirk of theology.

Ameroindian culture had already intuited these formative forces. The *Popol vuh*, the Maya bible of the Americas, has Four Creations which dovetail with the quadruple pattern of Mesoamerican cosmogony. Here the Seven Parrot drama has "its fourfold logic, which admits of no subtraction or addition" and when completed stands "perfected."[36] The Fourth World's best known icon, the opening map of *Féjérváry* has four major lobes or quarters with quatrefoil prevailing:

"As in the *Coixtlahuaca Map* [the] opening out of the directions is accompanied by an argument through embedded or subsidiary numeracy, here biological and botanical rather than geological. Hence, we find that the flowering trees of south and north constitute a pair. The former triples the twice double three of the Eastern tree (i.e. twice $3 \times 3 \times 3 = 54$ leaves); the latter modifies this in the name of calendrics, reducing the 54 to 52 year leaves of the Round (i.e. twice $3 \times 3 \times 3$ less $2 = 52$ leaves). Put another way, the relentless base arithmetic of the underworld with its squares and cubes is fitted to the more idiosyncratic phases of the sky as these are regulated in the lunar *tonalamatl* and the solar year Signs (4) and Numbers (13; $4 \times 13 = 52$). Like the subsidiary numeracy that defines the height of the judges' thrones in *Mendoza*, this persistent distinction in the *Coixtlahuaca Map* and *Féjérváry* between upper and lower arithmetical modes may possibly have specific social and class resonances."[37]

Many of the significances of these genuine mysteries have still to be unravelled. The subtle exactitudes of the Ameroindian systems remain, in their scope and extension, quite without parallel, a unique fusion of imagination and proto-science.

Mexican art emphasizes this. When the Great Coatlicue was dug up in Central Square, Mexico City, it was sent to be displayed at the University. But its effect proved so awesome that the city reburied it. As Octavio Paz explains, this

"is a repetition, on a reduced scale, of what the European mind must have experienced at the time of the discovery of America. The new lands appeared as an unknown dimension of reality. The Old World was ruled by the logic of the triad: three times, three ages, three persons, three continents. America had literally no place in this triadic world view. After its discovery the triad was dethroned. No longer were there just three dimensions and only one true reality; America represented a new dimension, and unknown fourth dimension. This new dimension was not governed by the principle of the triad but by the number four. The American Indians considered space and time, or rather space/time to be a single and dual dimension of reality that corresponded to the order established by the four cardinal points: four destinies, four gods, four colors, four ages, four worlds beyond the grave. Each god had four aspects; each space, four directions; each reality four faces. The fourth continent had emerged as a full and palpable presence. . . . It was not another reality, but rather the other facet, the other dimension of reality."[38]

So geography and intuitive thought preceded in their precision the four-dimensional space-time of Einstein, the journeying around the curved globe offering a foresight of centuries later theoretic discoveries.

NOTES

1. Hopper, 83–84.
2. Westcott, 50.
3. Christopher Butler (2), 3.
4. Farbridge, 118.
5. Schopenhauer (1), 3:325.
6. Christopher Butler (1), 19.
7. Dawkins (2), 110.
8. Turnell, 25.
9. Reiss, 161.
10. Beckwith, 254.
11. Jung (2), 73–74
12. Jung (1), 547–48.
13. Röstvig, 56.
14. Ovid, 347–41.
15. Burrow, 12.
16. Southern, 165.
17. Byrhtferth, 12–14.
18. Byrhtferth, 14.
19. Byrhtferth, 201–203.
20. *The Kingis Quair*, 48.
21. Kratzmann, 47.
22. MacQueen, 96.
23. Röstvig, 64.
24. Fowler, 33, 24.
25. Fowler, 156.
26. MacQueen, 96.

27. Berry, 148.
28. Dante, 29.
29. Danter, 87–81.
30. Steiner (3), 74.
31. Duckworth, 187.
32. David Simpson, 144
33. Guthrie, I, 2.
34. Graves, 37–38.
35. Lawrence (1), 137–33.
36. Brotherston, 225.
37. Brotherston, 107–01.
38. Paz, 18.

Chapter Fourteen

The Spread of the Fourfold Constant Into Nineteenth and Twentieth Century American Literature

Indirectly this continues into modern American literature. Here the matter becomes considerably more complex, and we have to dig out how the numbers are working. But by trusting the tale rather than the artist, *Moby-Dick*, for all its gigantic apparent formlessness yields a very clear mathematics. "I have written a wicked book" Melville announced to Hawthorne, and from this remark Henry A. Murray drew the inference that "all interpretations which fail to show that *Moby-Dick* is, in some sense, wicked have missed the author's avowed intention."[1] Melville wondered, not entirely ironically, whether "my evil art raised this monster" on reading Evert Duyckinck's cutting that reported the sinking of the *Ann Alexander*, having a few months earlier offered Hawthorne the tentative clue that "this is the book's motto (the secret one),— Ego non baptiso te in nomine—but make out the rest yourself."[2] In releasing "certain wondrous occult properties" Melville drew on ancient numerical and symbolic traditions to break up the complaisances of the novel of moral character.[3] In the process, he transforms plot rooted in time and Aristotelian sequence into one cast on an abstractly mathematical symmetry. His creation contains its deconstructive opposite while his archaism negates its modernism, negation here signifying the simultaneous inclusion yet surpassing of the modernist element by virtue of an Eliadean 'eternal return.'

Melville wrote in 1849 "the Declaration of Independence makes a difference,"[4] and *Moby-Dick* marginalizes the European conception of plot even as it reaches its apogee in Tolstoy, Balzac, Manzoni and Dickens. European narrative is, by implication, exposed as the artistic expression of the Old World plot to maintain time and the order of legitimate succession intact. Melville's transmutation of the significance of action and his dissolution of conventional time is not effected solely by biblical archetype and eschatology, nor only by the theological ambiguities drawn out by Lawrance Thompson in *Melville's Quarrel with God*. Rather it is that number ultimately determines identity and event in the novel, just as it is the key to nature's transcendence of "accidental and individual symbols."[5] It also rescues life from "the endlessness, yea, the intolerableness of all earthly effort."[6] But the rescue is dramatic and hard won, with number ripped from the flux of reality in contradistinction to an Emersonian a priori presence.

If in progressing from *The Scarlet Letter* to *Moby-Dick* "we move from the Newtonian world-as-machine to the Darwinian world-as-organism" as Walter Bezanson argues, then Melville's apparently loose baggy organism has a clear structure even if it is not neo-classical.[7] Indeed as Ahab chases Moby around the globe, we seem to have moved to the warp of Einstein's relative spacetime. The tension between the work's affinities with the cosmic order and Melville's residual Calvinism which instilled a right sense of sin in being so attuned to nature, results in the number symbolism burying itself deep within the novel. Moreover the conception of nature as a token of God's perfection was shared with Goethe, Carlyle and Emerson, so that in confronting unflinchingly and uncompromisingly "a part of the universal problem of things" whereby sharks "contrive to gouge out such symmetrical mouthfuls (293)", Melville is pitting his thorough interpretation against his powerful older contemporaries.

While the protagonists of *Moby-Dick* are typologically cast, then from the beginning number also rivets them to their fate. The ocean into which the *Pequod* sails is a fusion of the threefold and fourfold, an objective numerical arena finding a patterned reflex in Queequeg's counterpane and, by implication, in his tattoos and coffin. This arithmetical character determines the domain outwith that directly controlled by humans. So Ishmael confides it was "those stage managers, the fates, put me down for this shabby part of a whaling voyage." The surveillance of his actions by the Three Morae "being cunningly presented to me under various disguises, induced me to set about performing the part I did." Under this triple influence he was driven to "the delusion that it was a choice resulting from my own unbiassed freewill and discriminating judgement (7)." The chimerical independence is parallel to that of the novelist himself.

In 'The Spouter-Inn,' the dialectics of triple and quadruple are laid out in the epitome of the painting. On entering the gable-ended inn, Ishmael finds himself looking upon a "thoroughly besmoked (12)" picture. This is a miniature of the novel and the reader's relation to it, a reader still in the process of being critically coaxed out by Melville. Triadic clusters of adjectives describe the painting's murky sense, a shaft of possible illumination only hinted in the quadruple, shrouded in doubt as it is almost immediately:

"But what most puzzled and confounded you was a long, limber, portentous, black mass of something hovering in the centre of the picture over three blue, dim, perpendicular lines floating in a nameless yeast. A boggy, soggy, squitchy picture truly, enough to drive a nervous man distracted. Yet there was a sort of indefinite, half-attained, unimaginable sublimity about it that fairly froze you to it, till you involuntarily took an oath with yourself to find out what that marvellous painting meant. Ever and anon a bright, but, alas, deceptive idea would dart you through.—It's the Black Sea in a midnight gale.—It's the unnatural combat of the four primal elements (10–11)."

The ambiguity of "black mass"—not merely adjective plus noun, but an adjectival noun also—bears a prognostic connection to Ahab's manicheanism. Ishmael's search for the painting's significance is undertaken "involuntarily" since any occult properties in humanity reside "hiddenly," and the whale-hunt has clear analogies with how the author understands perception to occur, how harpoon-like a "deceptive idea would dart you through." Interpretation is a matter of intuitive and inspirational insight, and enveloped in dangers as great as those posed by the whale to Ahab and his crew. Melville is already rendering the familiar inscrutable, and transforming the customary world into a metaphysical ocean.

Ishmael decides the painting depicts a trinary disaster. It portrays "a Cape-Horner in a great hurricane" with the sinking ship's "three dismantled masts alone visible," and the whale "in the enormous act of impaling himself upon the three mast-heads (13)." This anticipates the climactic chase where "the critical third day (566)" brings Ahab's open clash with Moby. In his *Studies in Classic American Literature*, D.H. Lawrence had dwelt, more analytically in the first than the final version of his chapter on *Moby-Dick*, upon the three mates aboard the *Pequod*. Starbuck, Stubb and Flask "symbolise the three parts of the psyche, reason, impulsive passion, and blind will. But all these three elements are subject to the monomaniac in Ahab."[8] And Leslie Fiedler uncovered an even richer seam of number in the organization of the crew, pointing out that Queequeg and Fedellah "are associated with two other representatives of the primitive, also harpooners, the Indian, Tashtego, and the African, Daggoo, Yellow, brown, red, and black, they seem, considered together, rather emblems than characters, signifying the four quarters of the globe and the four elements as well, for Daggo is carefully identified with the earth, Tashtego with the air, the Parsee with fire, and Queequeg with water."[9] So the "unnatural combat of the four primal elements" in "the black mass" of the painting at Spouter Inn is embodied in the crew.

If the painting in the Spouter-Inn prefigures the dimensions of the novel as whole, then Queequeg's counterpane contains a series of signs that hold out a promise of later salvation. It is a marriage of the quadruple and triple, echoing the form of the inscriptions on the Polynesian harpooner's body, "a complete treatise on the art of attaining truth . . . whose mysteries not even himself could read (480–81)." The counterpane itself is "patchwork, full of odd little parti-colored squares and triangles," which are of a piece with his "squares of tattooing (25,21)." Ishmael's quizzical attitude toward the harpooner mixed with a degree of patronisingness, has an innocence bordering on prejudice. It is his misapprehension that "Queequeg, do you see, was a creature in the transition state—neither caterpillar nor butterfly. He was just civilized enough to show off his outlandishness in the strangest possible manner. His education was not yet completed (27)." Yet to provide the coffin that will allow the Christian his survival, the cannibal must have concluded his holometabolous process or the obscure hieroglyphs are reduced to the merest pupatory green-horness.

Possibly though we are to interpret the coffin as just such a chrysalis, one that can deliver an ingenuous Ishmael to tell the tale of the whale-hunt as he is doing throughout the book, from his newly-minted condition, a hybrid now of civilized and primitive, Christian and partly Islamicized. Melville gives the ironic clue to Ishmael's callow misreading of Queequeg in the word "outlandishness," a quality that in its rough and ready maritime character already suggests a transformation beyond that of a mere landlubber like Father Mapple who can manage only a vegetative organicism, as self-renewal like "spring verdure peeping forth even beneath February's snow (38)." There is also the inference that the rhetoric and mock-seriousness of his sermon on the whale arises from this land-locked limitation. It is significant that Queequeg is an assiduous observer of Ramadan, and is not just a natural pagan. The philosophy of self-denial and self-abnegation so crucial to the redeeming coffin is at the heart of Islam, a word that actually translates as resignation or surrendering. Only Queequeg truly yields to the laws of nature; in the end, the residual Christian is saved not only by Jehovah but effectually by Allah as well. Although Ishmael carries many Western misconceptions, nonetheless as the traditional ancestor of the Arabs, he is partially reclaimed for Islam by the end of *Moby-Dick*. If "Queequeg becomes virtually a double, a shadow self for Ishmael" as Carolyn Porter argues,[10] then by the same token the Muslim shadows the Judeo-Christian. So the narrator of the novel who has already assimilated the experiences he is relating, has earned his representative name and, drawing on his lineage not only by way of the Old Testament, but also the Koran, can now entreat us to "Call Me Ishmael" in its full theological duality.

Melville's awareness of numerical determinants is clear from "A Bosom Friend." Here Ishmael breaks in upon Queequeg with "that little negro idol of his," and being thus disrupted he "going to the table, took up a large book there, and placing it on his lap began counting pages with deliberate regularity; at every fiftieth page—as I fancied—stopping for a moment, looking vacantly around him, and giving utterance to a long-drawn gurgling whistle of astonishment. He would then begin again at the next fifty; seeming to commence at number one each time, as though he could not count more than fifty, and it was only by such a large number of fifties being found together, that his astonishment at the multitude of pages was excited (49)."

The significance of numbers in this rite may indeed be "fancied." But the apparently random events of the *Pequod*'s voyage have a more deliberated structure. From the beginning the ship sails under the auspices of the tridimensional, her masts standing "stiffly up like the spines of the three old kings of Cologne" to face the challenge of "all four oceans (69)." Captain Peleg sits in "a strange sort of tent, or rather wigwam," which is "of a conical shape," with "a triangular opening (70)." The whale-hunt is to be three years in duration, while even the ship's white cedar banks are of "full three per cent (105)." And in an ill-omened echo of Ishmael's metaphorical Queequeg—"neither caterpillar nor butterfly"—the *Pequod*'s crew sets sail in the knowledge that should shipwreck occur, they are choosing "to perish in that howling infinite" having dispensed with security, "for worm-like, then, oh!, who would craven crawl to land (107)."

If Ishmael's angle of vision conceived Queequeg to be in a transitional, chrysalitic state, then Ahab remains at a far more embryonic stage. "'D'ye mark him, flask?' Whispered Stubb; 'the chick that's in him pecks the shell. T'will soon be out' (160)." Ahab's harpoon is, in its triplicity, to prove insufficient to the task: "pole, iron, and rope—like the Three Fates—remain inseparable (490)." And there is the suggestion that Ahab's diabolism is too nearly analogous to the trinal ship the *Bachelor* to succeed. So the fire rite in "The Candles" for all its savage fury, goes no farther and strikes no deeper than that carefree ship with "three men at her mast-head" and "three Long Island negroes . . . presiding over the hilarious jig (493–94)." Here in Ahab's pagan ritual, the trinacarian has formed the centre of proceedings, the yardarms "'touched at each tri-pointed lightning rod-end with three tapering white flames, each of the three tall masts . . . silently burning in that sulphurous air, like three gigantic wax tapers before an altar (505)." Now "Daggoo loomed up to thrice his real stature," and the three masts are metamorphosed to a promissory three spermaceti candles, while *the nine flames leap lengthwise to thrice their previous height (506–07)."

There is a chapter in *Mardi*, "They are Becalmed," in which a preternaturally absolute and suffocating stillness descends on the ship's progress. The immobility has "lasted four days and four nights."[11] But as usual in mature Melville, the moment soon becomes metaphysical, transcendental in a different sense from Emerson's intimations of God's designs, in being a product of the very imperfectness of experience. If a symmetry is to be found, it has to be salvaged from the ostensibly indifferent and meaningless:

> "On the third day a change came over us. We relinquished bathing, the exertion taxing us too much. Sullenly we laid ourselves down; turned our backs to each other; and were impatient of the slightest casual touch of our persons. . . .
>   The Four days passed. And on the morning of the fifth, thanks be to Heaven, there came a breeze."[12]

The experience moves inwards to reveal a mental core, a nucleus of the dialectic of the three- and fourfold within the context of a book that is Melville's third, taking its name from the third day of the week, the third month and the third planet."[13] But now setting aside the almanac and astrology, he moves on without apparent compass to the inchoate tempest of *Moby-Dick*.

However Melville's metaphysical rage does have an inner form, a type of instinctual mathematics. Ahab's wife is "not three voyages wedded," while his wound sustained off Cape Horn left him lying "like dead for three days and nights (79,92)." It was at the age of forty that he was branded, a significant biblical multiple of four. Sitting forebodingly on "that tripod of bones," he seems to Stubb to be "that cursed pyramid," related in its absoluteness to Moby itself with its "high pyramidical white hump" and which appears during the final chase "to have treble-banked his every fin (129, 131, 183, 552)." The Pythagorean ultimateness of Moby in its silence defies argument and understanding, proclaiming the brute force of identity outside God's covenant. Carolyn Porter has pointed to the reiteration of the threefold in Chapter 45:

> "Ishmael claims to have 'personally known three instances' in which a whale has been harpooned, escaped, and 'been struck again by the same hand.' But instead of enumerating each instance in orderly fashion he confuses them to the point of absurdity. He seems particularly obsessed, for example, with the 'three year instance,' in part no doubt because it echoes, and seems to underscore, his opening claim that he knows of three instances of a harpooned whale being reharpooned by the same man. The number three is repeated nine times in the course of one paragraph. . . . But the more Ishmael tries to make his primary point—that such encounters are not so unlikely as they seem—the more confused his argument becomes. Digression gives way to the simple and repeated insistence, 'I say, I, myself, have known three instances. . . .' The reductio ad absurdum arrives with his claim that in the three year instance he was 'on the boat, both times,' and recognized the whale by the 'huge mole' under its eye."

Porter concludes that Ishmael's "peculiar handling of his evidence obviously results from a deliberate strategy."[14] Cast as it is in the form of legal, Melville is having Ishmael breaking down the legalistic formalism to suggest a stuttering undertow of numerical significances but ones which are hardly susceptible to the linear expression of reasonable discourse.

The architecture of the sperm whale involves a natural intermeshing of the triple and quadruple, "forty and odd vertebrae in all" which "mostly lie like the great knobbed blocks of a Gothic spire, forming solid courses of heavy masonry. The largest, a middle one, is in width something less than three feet, and in depth more than four (454)." This is somewhat akin to the Lincoln Cathedral of *The Rainbow* in which Will and Anna Bragwen fight out their responses to its structures. The primary difference is that Melville's edifice is embedded in the nature of things and is a player in a drama more thoroughly cerebral. If D.H. Lawrence preaches a gospel in which "the soul is neither the body nor the spirit, but the central flame that burns between the two," Melville adheres to a more archaic concept of the head and seminal fluid as interdependent, the Greek soul prior to the divorce between mind and male body.[15] Moby's vertebrae refract to Ahab his experience of forty years in a wilderness of pillaging nature, culminating in the mission to wreak a final vengeance on Moby: "Forty-forty-forty years ago! Forty years of continual whaling! forty years of privation and peril, and storm-time! forty years on the pitiless sea! for forty years has Ahab forsaken the peaceful land, for forty years to make war on the horrors of the deep! Aye and yes, Starbuck, out of those forty years I have not spent three ashore (543)." As he strikes out into the ocean of fourfold memory, the threefold holds out the olive branch of life ashore, the number of course being traditionally associated with the goddesses of earth. And so microcosm and macrocosm move to their point of tragic bisection, the captain shatters "the magic glass (544)" of the common familial and personal humanity he shares for a moment with Starbuck before the final assault.

The Faustian aspect of *Moby-Dick* is most intense in "The Doubloon" whose "strange figures and inscriptions" Ahab set to interpreting "in some monomaniac way (430)." The doubloon may by definition of value be only twofold, but it bears four letters and four likenesses. The three Andean summits are overarched by "a segment of the partitioned zodiac" and it has "a dark valley between three mighty, heaven-abiding peaks, that almost seem the Trinity, in some faint earthly symbol (431–32)." In fact it embodies that dramatic projection of the dialectic of triple and quadruple determining Ahab's destiny. Beside Moby's head which has "a certain mathematical symmetry (329)," it is seen to be a melodramatic representation fitting for Ahab. Both these structures go well beyond the Carpenter's workmanlike and pure concept of "none but clean, virgin, fair-and-square mathematical jobs, something that regularly begins at the beginning, and is at the middle when midway, and comes to an end at the conclusion (525)." The coffin he transofrms into a life-buoy proves too thirled to triplicity to save: "I'll have me thiry separate, Turk's-headed life-lines, each three feet long hanging all round to the coffin (526)." Earlier after encountering the *Virgin* when a lesser whale is hunted by three boats held by three threads tied to three bits of board, mathematical limitations did not matter. For this was not the crucial prize, but merely an old humped bull whale. There ensues an orgy of easy triplicate success, "all three tigers—Queequeg, Tashtego, Daggoo" armed with "their three Nantucket irons (355)" able to effect a straight-forward kill.

The Bible supplies the ultimate quadruplicity. Ishmael's account of his survival is prefaced by the repeated sentence of the four messengers of Job's travails. This completes the Gaudíesque baroque arch begun by the Book of Jonah which, as Father Mapple's sermon had announced, consists of "only four chapters—four yarns (42)." Newton Arvin following Dante's "fourfold interpretation" and Augustine's *sententia* proposed the juxtaposition in *Moby-Dick* of the literal, oneiric, moral and mythic levels of meaning. And he distinguished a tetradic movement in the novel, "from one wave-crest to another." The opening chapters preceding the sailing of the *Pequod* are comparable, argues Arvin, to the first four books of *The Odyssey*, while the second wave-peak occurs when Ahab nails the doubloon to the main-mast:

> "The whole central portion of the book, from the sunset scene in Ahab's cabin to the encounter with the bitterly misnamed craft, the *Delight*, forms a third movement. . . . The fourth most naturally begins with 'The Symphony' and comes to a close with the catastrophe itself—the Epilogue forming a kind of musical coda, not wholly unlike the burning of Hector's body on the funeral pyre in the last few lines of the *Iliad*."[16]

Part Three, perhaps reflecting the power of the triform to impel the male hero unconsciously, mythically encapsulated in the Morae and expressed in Ahab's trinal litany of curses when the *Pequod* goes down in the three episodes of "The Chase," is by far the longest segment of the novel, occupying two-thirds of the whole.

Discrete number alone makes life significant, but Ishmael's rescue requires a salvatory intermeshing of the threefold and fourfold such as was found on Queequeg's counterpane, and by implication on his tattoed body and coffin. When Melville spins from the vortex of *Moby-Dick*, Pierre sets to work on his book with Plotinus Plinlimmon's pamphlet in 333 lectures as his starting-point in the search for "the talismanic secret, to reconcile the world with his own soul," without which for Pierre "there is no peace for him, no slightest truce for him in this life."[17] Despite his protestations to the contrary, Melville himself is almost crushed between the quest for an equation that will unlock the universe, and the "mutilated stumps" with which he has to make do.[18] Billy Budd's stutter will be a legacy of this, drawing the "upright barbarian" into the fatal error of doing violence to Claggart, the representative of Fallen knowledge and its tyrannies.[19]

Melville had remarked in a letter to Duyckinck in March 1849 that Shakespeare, writing in the age of absolute royal and ecclesiastical jurisdiction, had shared "the muzzle which all men wore over their souls" so that his dark insights had to be introduced surreptitiously. However he goes on to argue that while the dramatist "was not a frank man to the uttermost," neither could a mid-century American "*in this intolerant Universe*" (Melville's emphasis) afford to be.[20] Melville's censor is inner though, the product of an almost superstitious fear of doing anything to encourage the release of humanity's fundamental depravity. Hence the meanings of *Moby-Dick* that he could not admit, are as thoroughly embedded in his novel as Shakespeare's parabolic span in his plays, implicit to the point of invisibility.[21] Leviathan in its war with the land-beast Behemoth proclaims the continual threat of unrestrained barbarism to overturn civil society. Melville quotes the opening sentence of Hobbes's *Leviathan* in the Extracts preceding the novel, and it is a contradiction of this cetacean symbol of the State built into *Moby-Dick* that the concept of the present order, the state of things, is also the embodiment of naked strength. On the other hand, the "artificial man" of Hobbes's whale-State has its Melvillean point of reference in the language of Ishmael's "centerless multilicity," and the neutrality of the narrative—"colorless because it is all color . . . voiceless because he is all voices"—is challenged by the diabolic content of the numerical sub-text with which Queequeg is associated though not identical.[22] The reader is pulled by an undertow creating "this backward motion toward the source,/ Against the stream."[23] The arithmetical configurations of *Moby-Dick* are an almost instinctual outgrowth from the number system Melville early absorbed from certain books of the Bible, primarily the books of the Prophets and Revelation, reinforced by his reading of the frequently numerological *Narrative of Voyages and Travels* by Amaso Delano. But his remark to Hawthorne quoted earlier, suggests he intended a precarious stratagem designed to call forth, indeed to create, a new type of reader for America. So the author almost imperceptibly transforms subjective into objective, internal into external. By the end, he has capsized our certainties by intimating sources of their justifiable extinction, and by fictionally enacting their dissolution.

The quadruple psyche also tells many modern tales. So in William Faulkner's *Intruder in the Dust*, it is four coins that are the pivot of the young white boy's

education. It was "in the early winter four years ago" that he offered the black
Lucas Beauchamp cash in thanks for his kindness whereupon Lucas

> "standing with the slow hot blood as slow as minutes themselves up his neck
> and face, forever with his dumb hand open and on it the four shameful fragments
> of milled and minted dross. . . .
>   He turned without haste and walked steadily to the creek and drew the four
> coins from his pocket and threw them out into the water: and sleepless in bed
> that night he knew that the food had been not just the best Lucas had to offer but
> all he had to offer."[24]

The coins are then translated into a single abstract moralized coin, symbol
of the reification of human intercourse, and an epitome of the coinage of false
relations between humans:

> "Because there was the half-dollar. The actual sum was seventy cents of course and
> in four coins but he had long since during that first few fractions of a second trans-
> posed them into the one coin and one integer in mass and weight out of all propor-
> tion to its mere convertible value; there were times in fact when, the capacity of his
> spirit for regret or perhaps just the simple writhing or whatever it was at last spent
> for the moment and even quiescent, he would tell himself *At least I have the half-
> dollar, at least I have something* because now not only his mistake and its shame
> but its protagonist too—the man, the Negro, the room, the moment, the day itself—
> had annealed vanished into the round hard symbol of the coin."[25]

Not only are the four coins the focus of the moral conflict at the centre of the
novel, but also quaternity determines the swivel of the narrative. Its unravel-
ling turns upon four words:

> "'Going where with you?' Aleck Sander said.
> And he told him, harsh and bold, in four words:
> 'Dig up Vinnie Gowrie.'"[26]

There are three rescuers of Lucas Beauchamp, led by an old woman, since the
men are "too busy with facks" to overcome indigenous prejudice. The best to
be hoped in Faulkner's world is an altruism that sets aside mere commonsensical reasoning—

> "not one of the three of them dared think now, if they had done but one thing
> tonight it was at least to put all thought of ratiocination contemplation forever
> behind them; five miles from town and he would cross (probably Miss Haber-
> sham and Aleck Sander in the truck already had) the invisible surveyor's line
> which was the boundary of Beat Four: the notorious, the fabulous almost and
> certainly least of all did any of them dare think now, thinking how it was never

difficult for an outlander to do two things at once which Beat Four wouldn't like since Beat Four already in advance didn't like most of the things which people from town (and from most of the rest of the country too for that matter) did."[27]

At the very last in the plot, the power of calculation reasserts its blank self:

"'That makes it out,' [Lucas] said. 'Four bits of pennies. I was aiming to take them to the bank but you can save me the trip. You want to count um?'

'Yes,' his uncle said. 'But you're the one paying the money. You're the one to count them.'

'It's fifty of them,' Lucas said.

'This is business,' his uncle said. So Lucas unknotted his sack and dumped the pennies out on the desk and counted them one by one moving each one with his forefinger into the first small mass of dimes and nickels, counting aloud, then snapped the purse shut and put it back inside his coat and with the other hand shoved the whole mass of coins and crumpled bill across the table. . . .

'Now what?' His uncle said. 'What are you waiting for now?'

'My receipt,' said Lucas."[28]

The business of everyday commerce has settled again to overlay the altruism that momentarily broke convention and habit. *Intruder in the Dust* becomes flaccid over the last seventy pages after its essential resolution, for the tension disperses with the conclusion of the dialectics of 3 and 4, which is an echo of the gulf between woman and man. Faulkner seems to be suggesting that only when this dialectic is bared at moments of moral crisis does the full range of humanity come into play, and can visionary behaviour be released.

The more profoundly inwrought American novelist, Nathanael West, wrote four novels which follow a negative configuration of quadruplicity and its transformations. In *Balso Snell*, West like his protagonist searches for his hero, searches for his subject, and indeed for his very self. Then his second tale, *Miss Lonelyhearts* relates the downfall of the agony aunt who can no longer believe his own fatuous if well-intentioned optimistic advice and is reduced to a hypothetical person incapable of self-realization. After this Lemuel Pitkin's aspirations in *Cool Million* are chopped up bit by bit, just as his body is ultimately. He subsequently drifts into the Nazi Party to become a martyr for their cause. The pupa thrown up by the big city ends up a fascist dupe. In West's fourth and greatest book, the masses do achieve transformation—but it is only a transmutation into locusts, that is consumers. All that can be projected in a society red in tooth and claw is a faltering momentum of the great threshing machine of daily commerce, just as nature allows the occasional escapee:

"Homer was on the side of the flies. Whenever one of them, swinging too widely, would pass the cactus, he prayed silently for it to keep on going or turn back. If it lighted, he watched the lizard begin its stalk and held his breath until it had killed, hoping all the time that something would warn the fly. But no matter how much he wanted the fly to escape, he never thought of interfering, and was careful not to budge or make the slightest noise. Occasionally the lizard would miscalculate. When that happened Homer would laugh happily."[29]

If nature has invaded society by the mid-twentieth century as conventions based upon the rituals of religion and social stability break up, then, alas, the metamorphoses are negative.

NOTES

1. Murray, 66. The Melville announcement is from Davis, 142.
2. Melville (2), 140, 133.
3. Melville (1), 143
4. Melville (2), 80.
5. Matthiessen, 43.
6. Melville (4), 60. All page references in the body of the text are to this edition.
7. Bezanson, 669.
8. Armin Arnold, 240.
9. Fiedler, 530.
10. Porter, 81.
11. Melville (3), 48.
12. Melville (3), 49–50.
13. Maxine Moore, 11.
14. Porter, 94–96.
15. Lawrence (2), 389. A splendid account of the earliest phase of the Greek notion of mind and soul is given by Onians, 116–22.
16. Arvin, 167. The earlier quotations are from 157–58.
17. Melville (5), 208.
18. Melville (5), 141.
19. Melville (1), 16.
20. Melville (2), 80.
21. In the light of the numerical theme of this section, it is worth adding that Hawthorne's novelistic evolution likewise has a Pythagorean configuration, spontaneously adopting a tetradic sequence. A quartet of novels from *Fanshawe* to *The Blithedale Romance* is followed by a further quartet in a fragmented run, passing via his transitional unfinished *The Ancestral Footstep* to the final fourfold helix which is a fractured replication of the earlier one, and runs from *The Marble Faun* to *Doctor Grimshawe's Secret*.
22. Guetti, 110, 26.
23. "West-Running Brook." In Frost, 329.
24. Faulkner, 3 and 15–17.

25. Faulkner, 20–21.
26. Faulkner, 86.
27. Faulkner, 94.
28. Faulkner, 246–47.
29. West, 298.

Chapter Fifteen

Thoreau's Insect Science

Early on in his *Journal*, Henry David Thoreau refers to his "microscopic eye, (*PJ*1:56)"[1] at this point in his career (1838) conceived as a positive value. In his first published work, *Natural History of Massachusetts*, he had already interpreted this in a scientific formulation—"Entomology extends the limits of being in a new direction, so that I walk in nature with a sense of greater space and freedom. It suggests besides, that the universe is not rough-hewn, but perfect in its details." And in a thoroughly Nabokovian assertion he continued: "Nature will bear the closest inspection; she invites us to lay our eye level with the smallest leaf, and take an insect view of its plain," before the flourish of a characteristic early rhetorical question—"Who does not recall the shrill roll-call of the harvest fly? There were ears for these sounds long ago, as Anacreon's ode will show (*Exc*, 42–43)." Nine years later, Thoreau carries this a stage further in "the inchworm progress of a sentence"[2] where nature is thoroughly Gulliverized:

> "The white stems of the pines which reflected the weak light-standing thick & close together while their lower branches were gone, reminded me that the pines are only larger grasses which rise to a chaffy head & we the insects that crawl between them (*PJ*3:261)."

He was always ready to look at the cosmos—like that liberated slave of the ancient world, Aesop—from the angle of an insect, though at other times when zooming to long focus, he felt this was a limitation. As Lawrence Buell has it—"One reason *Walden* makes such strenuous reading is that every moment, or so one often feels, is made to seem the ultimate moment; every object is the transfiguration of itself; nothing, however small, is small."[3] And it is this dynamic perspective that makes his observations of insects so fresh, far fresher than many technically exact entomological records.

So in a passage from the *Journal* of 1857, "You turn over the whole vegetable mould, expose how many grubs, and put a new aspect on the face of
the earth. It comes pretty near to making a world (*J9*: 310–11)." The collo-
quialism accustoms the reader to the minute perspective, as though such ca-
sual activity were indeed the most natural thing in the world. In fact for
Thoreau 1857 is a key date in re-defining the world in the image of insects.
Three years on from the publication of *Walden* he has firmly re-adjusted his
perspective. In May he writes: "Is a house but a gall on the face of the earth,
a nidus which some insect has provided for its young? (*J9*:412)." A month
later he counters this parody of conventional domesticity with a transformatory metaphor about his own predicament: "Many creatures—devil's-needles,
etc., etc.—cast their sloughs now. Can't I? (*J9*:412)." Later in the year he
metamorphoses the generality of Marvell's "green thought in a green shade"
into a matter of entomological identification in one of its less usual signfi-
cances:

> "I see one of those peculiarly green locusts with long and slender legs on a grass
> stem, which are often concealed by their color. What green, herbaceous,
> graminivorous ideas he must have! I wish my thoughts were as *seasonable* as
> his! (*J10*:26)."

The mid-1850s was the time of an explosion in Thoreau's interest in insects,
and during this he filled out a profounder approximation of human life to the
natural world, and specifically the entomological. But it was also the period
of the death of his great entomological mentor, Thaddeus Harris in 1856.

Thoreau read and transcribed passages from Coleridge's *Hints towards the
Foundation of a More Comprehensive Theory of Life* with its idea that if nature "had proceeded no further, yet the whole vegetable, together with the
whole insect creation, would have formed within themselves an entire and independent system of life" when it first appeared in 1848.[4] His interest in insects started to develop soon after this, a slow gestation, though whether it
was a development from his observations of nature, or his reading is difficult
to say. Already by 1852 he had begun to take specific note if insects were absent, and in that year he pursued a particular interest in wasps, hornets and
bees. His first observation of a butterfly was 1850, while he developed a concern with caterpillars—or "worms" as he called them following the biblical
usage—during the years 1853–54. He became especially observant of co-
coons during 1854. In July 1856, he has honed his gaze to noting ova, those
of devil's needles—that is dragonflies—and of ants in August. Curiously in
1858, insects fade from sight, but then the 1859 *Journal* is full of them and
immediately in January he remarks that "This ice is a good field for entomologists (*J11*:42)." It is the first time he has spoken in these terms, even if

he is only obliquely referring to himself. It is almost as if 1858 had been a year of gestation, perhaps after an interim year's consideration of Harris's death. There is a sense of awakening not only of the new year and its seasons, but of his own consciousness. He comes to realize that what he thought were cracks in the ice were in reality caterpillars washed up by the freshets.

Now Thoreau is prepared to tease out insect identifications. What are the larvae of January 22, 1859, he wonders aloud, and decides they are the females of the lightning bug. He is sketching out a whole miniature vision of the world, in line with his Concord days, but containing more truth than any macrocosmic view. His strength as an entomologist is that he always sees the insect in its relation with its plants: "So quickly and surely does a bee find the earliest flower, as if he had slumbered all winter at the root of the plant. No matter what pains you take, probably—undoubtedly—an insect will have found the first flower before you (*J*7:320)."

Again later, Thoreau claims the creatures as the first entomologists in his naturally reconstituted society: "Observed some of those little hard galls on the high blueberry, pecked or eaten into by some bird (or *possibly* a mouse), for the little white grubs which lie curled up in them. What entomologists the birds are! Most men do not suspect there are grubs in them, and how secure the latter seem under these thick dry shells! (*J*8:116)."

By 1857 Thoreau approximates to a sociobiological view of society. He has become far more theoretically adventurous in his insect-society/economy speculations, hearing "at a distance the hum of bees from the bass with its drooping flowers at the Island, a few minutes only before sunset. It sounds like the rumbling of a distant train of cars (*J*8:416)." This is fleshed out when he involves his favoured insect—the cricket. Then there is a great leap of the imagination. He uses the world of the insect to launch a critique of capitalism, and specifically the enormous expansion of the late 1850s which was so disrupting all sectors of American society:

"The merchants and banks are suspending and failing all the country over, but not the sand-banks, solid and warm, and streaked with bloody blackberry vines. You may run upon them as much as you please,—even as the crickets do, and find their account in it.
   They are the stockholders in this bank, and I hear them creaking their content. You may see them on change any warmer hour. In these banks, too, and such as these, are my funds deposited, a fund of health and enjoyment. Their prosperity and happiness and, I trust mine do not depend on whether the New York banks suspend or no (*J*10:92–93)."

He ironically diminishes the mighty "threshing machine" of industry to the life of the common black field crickets:

"Thus they stand at the mouth of their burrows, in the warm pastures, near the close of the year, shuffling their wing-cases over each other (the males only), and produce this sharp but pleasant creaking sound,—helping to fetch the year about. Thus the sounds of human industry and activity—the roar of cannon, blasting of rocks, whistling of locomotives, rattling of carts, tinkering of artisans, and voices of men—may sound to some distant ear like an earth-song and the creaking of crickets (*J*10:107)."

The music of the Orthoptera in the Thoreauvian world-view assumes the part of the music of the spheres in medieval cosmology. For the creak of the cricket "affects our thoughts so favorably, imparting its own serenity" that "It is time now to bring our philosophy out of doors." The cold earth of the early year has become modified by "the spring cushion of a cricket's chirp," "nourishing our ears when we are most unconscious (*J*9:403)." Somewhat as for so many twentieth century South American poets, particularly Jorge Carrera Andrade, the cricket offers up "the earth-song (*PJ*3:137)."[5] But perhaps no writer has so integrated the sounds, perspective and life of the insect into an intellectual and imaginative philosophy as fully as Thoreau.

Even the water-bugs are the occasion for reflections on human mortality:

"They go to bed in another element. What a deep slumber must be theirs, and what dreams, down in the mud there! So the insect life is not withdrawn far off, but a warm sun would soon entice it forth. . . .

Ah, if I had no more sins to answer for than a water-bug! They are only the small water-bugs that I see. They are earlier in the spring and apparently hardier (*J*10:255–56)."

Thoreau never takes his eye off nature, even when moralizing, in his kaleidoscopic outlook. Later on, at the turn of the century, John Muir is far more practical and utilitarian. So Muir in a parallel description in *The Story of My Boyhood and Youth* writes:

"We were great admirers of the little black water-bugs. Their whole lives seemed to be play, skimming, swimming, swirling, and waltzing together in little groups on the edge of the lake and in the meadow springs, dancing to music we could never near. The long-legged skaters, too, seemed wonderful fellows, shuffling about on top of the water, with air-bubbles like little bladders tangled under their hairy feet; and we often wished that we also might be shod in the same way to enable us to skate on the lake in summer as well as in icy winter. Not less wonderful were the boatmen, swimming on their backs, pulling themselves along with a pair of oar-like legs."[6]

By the end of the nineteenth century, the battle to protect a diminishing natural environment has already become paramount. The insect vision is on behalf of the whole ecology, "on the preservation of the Sierra forests, and the wild and trampled conditions of our flora from a bee's point of view."[7] Muir's prose is more muscular and rugged than Thoreau's, sometimes merely cumbersome, but consistently polemical, with more insects as he gets older. He defines himself as "a strong butterfly full of sunshine [who] goes by crooked unanticipated paths from flower to flower."[8]

From a socio-political standpoint, Thoreau clearly equated the insect's place in the natural order of things with that to which the native Americans had been depressed. There is a suggestion that their language was nearer to nature than that of the New World settlers:

> "Our scientific names convey a very partial information only; they suggest certain thoughts only. It does not occur to me that there are other names for most of these objects, given by a people who stood between me and them, who had better senses than our race. . . . The Indian stood nearer to wild nature than we. . . . It was a new light when my guide gave me Indian names for things for which I had only scientific ones before. In proportion as I understood the language, I saw them from a new point of view. . . . The Indian's earthly life was as far off from us as heaven is (J10:294–95)."

This remarkable passage epitomizes Thoreau's attitude to the specialists. For he always expressed a liking for the work of "the old naturalists (J13:149)," and declares he has been "these forty years learning the language of these fields that I may the better express myself (J10:191)." This is inextricably bound up with the sense that "most scientific books which treat of animals leave this [the *animal*] out altogether, and what they describe are as it were phenomena of dead matter (J13:154)." Thoreau watches small grubs splitting a *Rhus toxicodendron* leaf, working to split the leaf but being "creatures so minute that they find food enough for them between two sides of a thin leaf without injuring the cuticle" (J12:308). His sheer fascination with this class of Animalia, however apparently ugly and ordinary they may be, is in part an inheritance from Jonathan Edwards when he marveled as a boy at the activities of a spider.

Spiders are not insects which are stubbornly six-legged. But they play such a micro-macro role in Thoreau's vision that some remarks on them are relevant. Sophistry, of course, has been imaged in spider/web terms from Plato to Pope and beyond, and there is something of this tradition—perhaps unconsciously—in Thoreau's imagery:

> "The grass is thickly strewn with white cobwebs, tents of night, which promise a fair day. I notice that they are thickest under the apple trees. Within the woods the most or dew on them is so very fine that they look smoke-like and dry, yet even there, if you put your finger under them and touch them, you take off the

dew and they become invisible. They are revealed by the dew, and perchance it is the dew and fog which they reveal which are the signs of fair weather (*J*:12:265–66)."

The cobwebs, "tents of the night" which yet portend "a fair day," a bringing to light, have yet the invisibility and concealed roots of real thought. He is fascinated by the "cobweb-tapestry (*J*:11:363)," finding that "Their geometry is very distinct (*PJ*5:195)"—"I see many perfectly geometrical cobwebs on the trees, with from twenty-six to thirty-odd rays, six inches to eighteen in diameter (*J*6:260)." Although Thoreau nowhere translates his observations into metaphorical terms, nonetheless insofar as we read the Journal as literature and not a mere catalogue of nature records, there can be discerned a parallel with writing as such in his description of "the fine lines" of "a gossamer day"—"They are so abundant that they seem to have been suddenly produced in the atmosphere by some chemistry, spun out of air,—I know not for what purpose. . . . These gossamer lines are not visible unless between you and the sun (*J*5:465)." Indeed they take on something of the expeditionary tackle of Ahab's Pequod:"They are not drawn taut, but curved downward in the middle, like the rigging of vessels,—the ropes which stretch from mast to mast,— as if the fleets of a thousand Lilliputian nations were collected one behind another under bare poles. But when we have floated a few feet further, and thrown the willow out of the sun's range, not a thread can be seen on it (*J*5:465–66)." So great writing is seamless, woven without stitches, and best read against the light of the sun. It is as if the writer (Thoreau's "year") "were weaving her shroud out of light." And although as on other occasions Thoreau refers to the spider as "the insect" with its "*point d'appui*," it presents a certain "magic" in this gratuitous, and apparently futile, stream of lines (*J*5:466).[9] To the good citizens of Concord, he was a saunterer. Not a parasite for he was a surveyor who could also manufacture the best of pencils. But his writing had something of the free extension of the spider's web so late in the season that "I saw no insects caught in it (*J*5:469)."

Much has been made of the spider's web in *Walden* by Lee Quinby in a subtle and Foucaultian argument. He interprets the spider/web imagery in that book as a stage in Thoreau's quest for chastity. In line with present day taboos of expression, Quinby claims hierarchy is at work in insect metamorphosis where successive steps are abandoned "in order to reach a higher one," while in the spider/web trope "all strands of knowledge and experience are equally vital, all meeting or supporting the others."[10] However this is a misunderstanding of the helical strands of imago production which, from the initial egg and larva stages, are contained in the recurrent imaginal buds. Metamorphosis is no more hierarchical than DNA when likewise the genetic material itself remains unchanged during the process of the development of the individual organism, and the different forms are simply the result of a complex program of information. And while it is true, as Quinby remarks, that summer, not spring, actually concludes *Walden*, this simply presages the picaresque records of his "continued rambling"[11] in his *Journal* as opposed to the created object that was his earlier book. Once Thoreau had defined himself in relation to nature's processes in *Walden*, he could move on to surprise the ecology of Concord.

There is a most strange passage in *Walden* where in what James McIntosh calls "a stream of consciousness,"[12] and what we might distinguish as intimations of a more Carlylean or De Quinceyan mode of writing, the author remarks: "Only a woodpecker tapping. O, they swarm; the sun is too warm there; they are born too far into life for me (*W*:223)." This is part of his progress during the pivotal year 1851 when his observations of flora and fauna become far more specific and exacting. This progress is closely bound up with his literary interests, for it is significant that entries in his *Journal* at the inception of his *annus mirabilis* centre on Ovid's *Metamorphoses*, most particularly regarding the transformations of Clymene's daughters into trees weeping amber tears on the death of Phaeton. In February that year he wrote that "My desire for knowledge is intermittent but my desire to commune with the spirit of the universe—to be intoxicated even with the fumes, call it, of that divine nectar—to bear my head through atmospheres and over heights unknown to my feet—is perennial & constant (*PJ*3:185)." By September though, he suggests the observing human brings life to nature, fertilizes it, and leads to a conceiving, perhaps ultimately to conceptions:

> "How to live. How to get the most life! as if you were to teach the young hunter how to entrap his game. How to extract its honey from the flower of the world. That is my everyday business. I am as busy as a bee about it. I ramble over all fields on that errand and am never so happy as when I feel myself heavy with honey & wax. I am like a bee searching the livelong day for the sweets of nature. Do I not impregnate & intermix the flowers produce rare & finer varieties by transferring my eyes from one to another? (*PJ*4:53)."

This relates in turn to his early angst with regard to the human place within nature, "How is it that man always feels like an interloper in nature, as if he had intruded on the domains of bird and beast (*PJ*1:398)." Much of the *Journal* seems directed to proving his independence from Emerson, and an essential aspect was his search for a meaningful philosophy that surpassed Emerson's idealist "analogy running through all parts of nature, & the correspondence of physics and metaphysics."[13] Part, at least, of the answer he found in the Cowperish oddity, globosity, of the gall which he characteristically extended far beyond the English poet's weighty factuality into a raking metaphor for nature itself:

"Saw some green galls on a goldenrod ('?) '? an inch in diameter shaped like a fruit or an Eastern temple with 2 or 3 little worms inside-completely changing the destiny of the plant-showing the intimate relation between animal & vegetable life—The animal signifies its wishes by a touch & the plant instead of going on to blossom & bear its normal fruit devotes itself to the service of the insect—& becomes its cradle & food. It suggests that Nature is a kind of gall— that the creator stung her & man is the grub—she is destined to house & feed. The plant rounds off & paints the gall with as much care & love as its own flower & fruit-adorning it perchance even more (*PJ*:6:282)."

Thoreau thus abandons analogy for a full-developed metaphorical and materialist interpretation of God-in-Nature.

Indeed he fearlessly disputed with the botanist Asa Gray a year earlier in order to set up his hypothesis: "Gray refers the cone-like excrescences on the ends of the willow twigs to the punctures of insects. I think that both these & the galls of the oak &c are to be regarded as something more normal than this implies. Though it is impossible to draw the line between disease & health at last (*PJ*4:463)." So in this brilliantly imaginative metaphor, Thoreau presents his definition of the human place in nature. Art is directly implicated in this, for "Is not Art itself a gall? Nature is stung by God and the seed of man planted in her. The artist changes the direction of Nature and makes her grow according to his idea (*J*7:10)." Joseph Sanborn Wade summarizes the conception—"It suggests that Nature is a kind of gall, that the Creator stung her and man is the grub she is destined to house and feed."[14] And so the idea of the artist can be juxtaposed entirely naturally with that of the insect. Observations can merge with philosophical reflections: "The butterflies are now more numerous red and blue-black or dark velvety. The art of life—of a poet's life is—not having anything to do, to do something (*PJ*5:8)." In that sense, the artist must emulate his or her natural cousins, take on something of the camouflage of nature, even its protective mimicry, but more positively "the *non chalance* of the butterfly carrying accident and change in a thousand hues upon his wings (*PJ*1:167)."[15]

NOTES

1. The abbreviations for the Thoreau titles embedded in the references within this chapter, and amplified in the Bibliography, are—
    PJ: volumes in the Princeton University Press edition of Thoreau's *Journals*
    J: the Torrey-Allen edition of Thoreau's *Journals*
    Exc: *Excursions*
    W: the Princeton University Press edition of *Walden*
2. Boyd, 210.
3. Buell, 305.

4. Coleridge, 11: i, 542.

5. Spooner (2), 119–21.

6. John Muir (2), 53.

7. Cohen, 230–31.

8. John Muir (1), 2.

9. For Thoreau's assumption that spiders are technically insects, see for example *J*9:120 and *J*13:42. For an excellent discussion of the spider's web in American literature, see Tony Tanner, *Scenes of nature, signs of men* (Cambridge: Cambridge University Press, 1987), 25–39.

10. Quinby, 76.

11. Quinby, 101.

12. McIntosh, 248.

13. Porte, 457.

14. Wade, 14.

15. The *Journal* gave Thoreau confidence to experiment further linguistically. In *Excursions*, "*non-chalance*" is printed straightforwardly as "nonchalance (40)."

Chapter Sixteen

The Scientific Ambiguities of Literary Creation: Nabokov's Art of Evolution

Vladimir Nabokov could have agreed with Stephen Jay Gould that criteria other than fitness for reproduction motored evolution. Nabokov was working at the Museum of Comparative Zoology at Harvard University between 1942 and 1948 at a time when Darwinian theory had not yet been fully adopted by evolutionists. Hence he felt free to speculate in a way that undoubtedly fueled his imaginative writing, while incurring the skepticism of today's evolutionary biologists. The father in *The Gift* enunciates to his son a perspective on mimicry in nature that is close to Nabokov's own—

> " He told me about the incredible artistic wit of mimetic disguise, which was not explainable by the struggle for existence (the rough haste of evolution's unskilled forces), was too refined for the mere deceiving of accidental predators, feathered, scaled and otherwise, . . . and seemed to have been invented by some waggish artist precisely for the intelligent eyes of man (a hypothesis that may lead far an evolutionist who observes apes feeding on butterflies); he told me about the magic masks of mimicry; about the enormous moth which in a state of repose assumes the image of a snake looking at you (*GIFT* 105.)" [1]

The Darwinians are confident that they have abolished the teleological aspect of evolution, but of course reproductive fitness retains such an aspect however random its functioning and determining its law. For Nabokov, these ideas in *The Gift* are a source of all his mature work. Brian Boyd surmises that he had misgivings about combining metaphysical speculation with proven scientific data in *The Gift*,[2] but his imagination was fertilized by such hypothesizing. It is precisely the contradiction between the scientific theory of natural selection and the freedom of speculation, that releases the imago of the fictional and biographical style. The luminosity of Nabokov's stories and novels creates a ghostly butterfly-cum-moth glow hovering over the fiction, what in another context he called "the bright mental image...conjured up by a wing-stroke of the will (*SM* 20)," or "the eye-spot of his awakening" in *The Real Life of Sebastian Knight* (31). Nikolay Gogol had achieved this—"As in the scaling of insects the wonderful color effect may be due not to the pigment of the scales but to their position and refractive power, so Gogol's genius deals not in the intrinsic qualities of computable chemical matter (the 'real life' of literary critics) but in the mimetic capacities of the physical phenomena produced by almost intangible particles of recreated life (*NK* 56)." The effect is related to the natural phenomenon of mimicry in insects, for each time the observer locates such mimicry, "the mind's birth" is re-enacted in "the stab of wonder that accompanies the precise moment when, gazing at a tangle of twigs and leaves, one suddenly realizes that what had seemed a natural component of that tangle is a marvelously disguised insect or bird. (*SM* 233)." Such instances of revelation are determining because they are inseparably connected with a sense of time, an epiphany, and as such "quite different from the spatial world, which not only man but apes and butterflies can perceive (*SM* 11)." Nabokov is implying that in his writing, the reader is seamlessly experiencing the interrelation between mind and nature, between mind and insect. This is related to the moment of pure objectivity he sought at the close of a novel, and invokes a fictional artist who creates "a sensation of its world receding in the distance and stopping somewhere there, suspended afar like a picture in a picture: *The Artist's Studio* by Van Bock (*SO* 72–73)." For he is redefining *Homo sapiens* as preceded by *Homo poeticus*, who embodied "the riddle of the initial blossoming of man's mind (*SM* 233)."

It is as though the books written in his maturity are palimpsests, appearing at the apex of the helix that has spiraled upwards from the time and experience they began. (Nabokov referred to "the essential spirality of all things in their relation to time (*SM*, 215).") The distillation that is his fiction enacts in the verve and swerve of his writing the very creatures which were his obsession. This is an implicit quality for, as he says, "whenever I allude to butterflies in my novels , no matter how diligently I rework the stuff, it remains pale and false and does not really express what I want it to express—what, indeed, it can only express in the special scientific terms of my entomological papers." (*SO* 136) One paper, unfortunately lost, which survives in brief extracts in *Ada*, *The Gift* and *Speak Memory* was the 1942 "The Theory and Practice of Mimicry." A clue as to its drift comes from a significant passage in "Father's Butterflies" where Nabokov points to the incidence of mimicry where no protective imitation is at work at all. This is in the caterpillar of the Siberian Owlet moth (*Pseudodemas tschumarae*) which is found on the chumara plant:

"Its outline, its dorsal pattern, and the coloring of its fetlocks make it resemble precisely the downy, yellow, rusty-hued inflorescence of that shrub. The curious thing is that . . . the caterpillar appears only at summer's end, while the chumara blooms only in May, so that, against the dark green of the leaves, the caterpillar, uncircled by flowers, stands out in sharp contrast. The resulting impression (if one adheres to the illusory theory of 'protective mimicry') is that there has been a hitch in fulfilling the agreement, or that, at the last minute, nature defrauded one of the parties."[3]

Perhaps there is a sort of ongoing metempsychosis in his fiction, on the lines of that explained by Aleksey Ivanovich in *Mary*: "'our old life in Russia seems like something that happened before time began, something metaphysical or whatever you call it—that's not quite the word—yes, I know: metempsychosis.'" (25) In *Ada* a character expresses an aspiration related to this, one that prefigures James Joyce's coupling of incest and insect: "'If I could write,' mused Demon, 'I would describe, in too many words no doubt, how passionately, how incandescently, how incestuously—*c'est le mot*—art and science meet in an insect, in a thrush, in a thistle of that ducal bosquet' (*ADA* 436)." Demon's idealist and adolescent limitations are incorporated in this statement of intent; Nabokov would resolve the matter far more subtly and intrinsically.

Nabokov was especially interested in the process and roots of imitation in nature:

"The mysteries of mimicry had a special attraction for me. Its phenomena showed an artistic perfection usually associated with man-wrought things. . . . When a certain moth resembles a certain wasp in shape and colour, it also walks and moves its antennae in a waspish, unmothlike manner. When a butterfly has to look like a leaf, not only are all the details of a leaf beautifully rendered but markings mimicking grub-bored holes are generously thrown in (*SM* 94)."

He saw aspects of mimicry as illuminating an inadequacy in Darwinian natural selection since "a protective device was carried to a point of mimetic subtlety, exuberance, and luxury far in excess of a predator's power application. I discovered in nature the nonutilitarian delights I sought in art. Both were forms of magic, both were a game of intricate enchantment and deception (*SM* 95)." The process fascinated the author because it was so close to *mimesis*. So, as Vladimir Alexandrov observes, "Nabokov's textual patterns and intrusions into his fictional texts emerge as imitations of the otherworld's formative role with regard to man and nature: *the metaliterary is camouflage for a model of the metaphysical*."[4] There was a spine-tingle in producing on the cultural level something akin to what was happening in nature, but which was consciously guided by the author just as nature's patterns demand a higher consciousness. As Cavalli-Sforza and Feldman explain the process:

> "it is also important to emphasize the distinction between *mutation* in the biological and in the cultural processes. In the former, mutation is typically an error in the copy, which as a result is not identical to the master, or it is a chemical change in the master to be copied. In both cases *randomness* seems to be the/rule. On the other hand, in the cultural process, the change is not necessarily a copying error, but it can often be directed innovation, that is, innovation with a purpose, and might therefore appear to be nonrandom."[5]

Indeed Nabokov's project of an even now unpublished illustrated *Butterflies in Art* related to the objective scientific side of this issue. His aim was to look at the depiction of the minutiae of butterflies in the Old Masters, and consider whether "evolutionary change can be discerned in the pattern of a five-hundred-year-old wing? (*SO*, 168)." Nabokov's famous gratuitous dewdrop beyond utilitarian needs appearing on Blue's wings is itself technically part of the question, and he hoped the paintings would shed light "on the time taken for evolution; one thousand years could show some little change in trend (*SO* 169)."

The designs of some butterflies in developing protective camouflage that imitates poisonous butterflies has in recent research come to be interpreted as the result of a long-term process of genetic adaptation.[6] Nabokov telescoped this process, seeing it as an act of saltation. In this he followed an early authority on the subject, R.C. Punnett, who surmised that "the new character that differentiated one variety from another arises suddenly as a sport or mutation, not by gradual accretion of a vast number of intermediate forms."[7] Stephen Jay Gould expresses serious reservations about much of Nabokov's entomological theorizing, but here the two were close in their variations of punctuated equilibrium.[8] Victoria Alexander in her pioneering paper on Nabokov, mimicry and evolution adduces dramatic evidence to back the writer:

> "It so happens that butterfly wing pattern development is particularly sensitive to climatic conditions. Many species have wildly different summer or winter or wet- and dry-season forms. . . . Studies have shown that near-lethal high or low temperatures temporarily disrupt the activity of certain genes during development. . . . When this happens to an entire population, the previously hidden wild-card diversity can be suddenly revealed, and, in turn, natural selection can act on it."[9]

As Alexander argues "Although natural selection might *stabilize* a resemblance once it is found, selection alone could not *create* it. This, I argue, is the crux of Nabokov's dispute with the Darwinists of his day."[10] For there is nothing generous about the Darwinian universe, no scope for excess. Nor does Darwin offer any particular role to symmetry. Alexander observes of Nabokov as entomologist:

"One of Nabokov's specialities was describing the relative shapes and sizes of butterfly reproductive organs, the basic shape of which is triangular. Aberrant members of the species tend to be less symmetrical, but the 'main peaks of speciation' argued Nabokov, exhibit a 'convenient constant in the structural proportions,' conforming to an equilateral triangle. It seemed to him as if symmetry were a goal toward which species strove. Such were the kinds of arguments made by the morphologist-teleomechanists in the nineteenth century. Nabokov noted that this symmetry had no bearing on reproductive capability; thus, Darwinian natural selection could not be brought in to explain it. Nabokov supposed instead that some laws of biological form might contribute to this phenomenon."[11]

He saw key parts of Darwinian theory as merely solipsistic, not untrue but self-fulfilling—in that an organism must survive in order to prove its fitness. Therefore in order to understand the human condition, it became essential to ask metaphysical questions. In parallel with sudden changes in evolution, Nabokov believed that "consciousness may have developed as the result of perceptual and cognitive leaps, rather than as a result of prior organic evolution that creates the material base or potential for these leaps (*LL* 243)." The relation of music to Darwinian theory, and the lack of a true correlation discussed in Chapter 4, bears out Nabokov's approach.

There is a sense of irony throughout Nabokov's work which derives from the contradiction between humanity as a mammalian biped, and as a creature paradoxically fused with the apparently remoter, yet mentally more intimate, world of lepidoptera. So in *The Real Life of Sebastian Knight*, Madame Lecerf "thought it would be rather good fun to have him make love to her—because, you see, he looked so very intellectual, and it is always entertaining to see that kind of refined, distant—brainy fellow suddenly go on all fours and wag his tail. (148)" At the same time, there are other patterns of human life. Van in *Ada* mimics his sister endeavouring to absorb her qualities. That novel plays upon camouflage and insect life. So Ada's "cryptic silence (86)" when aroused by Van conjures up the cryptic markings disguising, say, a Comma butterfly among the scalloped leaves of an oak tree. Humans, like insects, utilize imitation for protection. Alfred Appel Jr. teases out one such pattern in his inspired notes to *Lolita*:

"One of Nabokov's lepidopterological finds is known as 'Nabokov's Wood-Nymph' . . . and he is not unaware that a 'nymph' is also defined as 'a pupa,' or 'the young of an insect undergoing incomplete metamorphosis.' Crucial to an understanding of *Lolita* is some sense of the various but simultaneous meta-morphoses undergone by Lolita, H.H., the book, the author, and the reader, who is manipulated by the novel's game-element and illusionistic devices to such an extent that he too can be said to become at certain moments, another of Vladimir Nabokov's creations—an experience which is bound to change him. The butter-fly is thus a controlling metaphor that enriches *Lolita* in a more fundamental and organic manner than, say, the *Odyssey* does Joyce's *Ulysses*. Just as the nymph undergoes a metamorphosis in becoming the butterfly, so everything in *Lolita* is constantly in the process of metamorphosis, including the novel itself—a set of 'notes' being compiled by an imprisoned man during a fifty-six-day period for possible use at his trial, emerging as a book after his death, and then only after it has passed through yet another stage, the nominal editorship of {the pioneer English entomologist} John Ray, Jr....[Humbert's] sense of a 'safely solipsized' Lolita is replaced by his awareness that she was his 'own creation' with 'no will, no consciousness—indeed, no life/ of her own, that he did not know her, and that their sexual intimacy only isolated him more completely from the helpless girl."[12]

Unfortunately Appel has fluffed the understanding of the entomological nymph through using the dictionary definition together with the confusing French equivalent ('nymphe') which signifies 'pupa.' He therefore misses the full irony and pathos of HH's problems. The point is that the nymph does not go through the decisive chrysalis stage, but merely matures through a series of ecdyses, a slow shedding of previous stages. It belongs to the hemimetabo-logical orders. Mary S. Gardner clarifies the matter in her *The Biology of In-vertebrates*:

> "Of the hemimetabolous orders, some, like mayflies and dragonflies, emerge from the egg as nymphs that do not bear close resemblance to adults . . . but their transition to the adult form, or imago, occurs in a series of steps during their juvenile life in which existing structures acquire adult form. Other hemimetabolous species, like cockroaches, grasshoppers, and bugs, hatch as nymphs that are essentially miniature adults, usually lacking only fully devel-oped wings, gonads, and external genitalia, and gradually acquiring these par-ticular adult features with successive molts."[13]

In other words the larval stage is much extended, misleading the observer as to the stage it has reached and so the moment of maturity is sudden. This lies behind Humbert's confusion—as well as his lust. One moment Lo was a nymph in his powers, but within a brief span of time she had become a do-mesticated wife. Rather strangely—but an aspect of Nabokov's suspicion of institutionalized and schematized intellectuals—Nabokov had no interest whatsoever in this type of symbolism: "That in some cases the butterfly symbolizes something (*e.g.*, Psyche) lies utterly outside my area of interest (*SO* 168)." He is though prepared to utilize such symbolism to the degree that Browning uses Canova's 'Psyche and Cupid' as a shadow drama in *Pippa Passes*. For him, the whole clanjamfry of symbolic and allegorical thought was a pedantic and computerized disease that "destroys plain intelligence as well as poetical sense. (*SO* 305)." More germane to Nabokov's intentions is his use of mimicry in *Lolita*. He had heralded this use in a poem "The Snapshot" of 1927: "I, the accidental spy,/I in the background have also been taken. . . . / My likeness among strangers,/one of my August days,/my shade they never noticed, / any shade they stole in vain. (*PP* 41, 43)" Humbert passes through middle consumer America unnoticed, his camouflage being his very seedy ordinariness as he moves from motel to motel with his quarry. A measure of the subtlety of Nabokov's novel can be gauged by comparing it with John Fowles's mechanical and one-paced *The Collector* published eight years after *Lolita*.

A Russian writer from Nabokov's youth, P.D. Uspensky, delved into the question of insect mimicry. In his *A New Model of the Universe*, he argues that such imitation explains "many things in the principles and methods of Nature."[14] Here Uspensky relates directly to the ideas of the dramatist and theatre director, Nicolas Evreinov in that both emphasize the theatricality of nature, its artificiality, a theme that appears in many of the fictions of Nabokov—the Red Admirable, for example, that alights on John Shade moments before his death, and that puts in an appearance in *King, Queen and Knave* on the disappearance of the fatidic narrator. As Alexandrov puts it— Nabokov shares with Uspensky and Evreinov "several unusual ideas, including the seminal redefinition of 'artifice' and 'nature' as synonyms for each other on the basis of mimicry among insects."[15] Uspensky though takes matters further, and poses the metaphysical-evolutionary riddle:

"But what place shall we give in this system to *insects*, which represent a world in themselves and a world not less complex than the world of invertebrates?

May it not be supposed that insects represent another line in the work of Nature, and live not connected with the one which resulted in the creation of man, but perhaps preceding it?

Passing to facts, we must admit that insects are in no way a stage preparatory to the formation of man. Nor could they be regarded as the by-product of human evolution. On the contrary, insects reveal, in their structure and in the structure of their separate parts and organs, forms which are often more perfect than those of man or animals. And we cannot help seeing that for certain forms of insect life which we observe there is no explanation without very complex hypotheses, which necessitate the recognition of a very rich past behind them and compel us to regard the present forms as degenerated forms. . . ."[16]

At some point, ants and bees lost their intelligence and ability to evolve and "after this Nature had to take her own measures and, after isolating them in a certain way, to begin a new experiment."[17] This is akin to a statement by a thinker far removed from Uspensky's perhaps excessively systematic philosophy, that is Coleridge's assertion quoted already that if nature "had proceeded no further, yet the whole vegetative, together with the whole insect creation, would have formed within themselves an entire and independent system of life."[18] Certainly Uspensky was prepared to go further than Nabokov in making his hypotheses precise, and as with symbolism and allegory, Nabokov was loath to commit himself to exact metaphysical projection. As Alexandrov expresses it: "Uspensky interprets mimicry among insects as pointing to the 'fourth dimension' (which is a concept that he did not invent, and which has a long and complex history in turn-of-the-century thought), and as evidence for an impulse in the cosmos whose aim is to produce a being capable of achieving transcendence."[19] Leaving aside some of the speculative evolutionary material ("degenerated forms"), it is true the world of the insect remains peripheral to contemporary thought. The obsession with human physiological links to the great apes has blinded theoreticians to an inner relation with the insects.

Uspensky sacrificed his own researches for discipleship of Gurdjieff. But his intimations may have remained at the back of Nabokov's own mind, though metamorphosed into a more general sense of patterning in nature. And to return to my earlier clumps of significant word-groupings, it is interesting that Nabokov saw mimicry as enmeshed with the foundation of language

> "in the tangle of sounds, the leopards of words,
> The leaflike insects, the eye-spotted birds
> fuse and form a silent, intense,
> mimetic pattern of perfect sense." (PP 157)

Bobbie Mason observed that "his kaleidoscopic spin of linguistic resemblances (puns, anagrams) [and I might add spoonerisms] . . . matches the phenomenon of mimicry in nature."[20] Much of the sense of anticipation, of excitement in reading Nabokov comes from this apprehension that a revelation is near at hand, some decisive clarification which will illuminate and in time change the human condition. Krug in *Bend Sinister* interprets consciousness as "the only real thing in the world and the greatest mystery of all," because "we live in a stocking which is in the process of being turned inside out, without our ever knowing for sure to what phase of the process our moment of consciousness corresponds (*BS* 156, 161)." In that sense, Nabokov's work is part itself an evolutionary mutation, but creatively directed rather than random. Here the Cosmic Cultural Faculty has taken the matter of evolutionary theory to a new precipice. Or as Nabokov puts it in terms bringing him tangential to Uspensky's concepts, to Einstein's relativity and to the most recent theories of the earliest stages of the cosmos: "every dimension presupposes a medium within which it can act, and if, in the spiral unwinding of things, space warps into something akin to time, and time, in turn, warps into something akin to thoughts then, surely, another dimension follows—a special Space maybe, not the old one, we trust, unless spirals become vicious circles again (*SM* 236)."

NOTES

1. Abbreviations used in the references embedded in the text:
   ADA: Ada or Ardor: A Family Chronicle
   BS: Bend Sinister
   GIFT: The Gift
   LL: Lectures on Literature
   NG: Nikolay Gogol
   PP: Poems and Problems
   SM: *Speak, Memory: An Autobiography Revisited*
   SO: Strong Opinions
   STORIES: The Stories of Vladimir Nabokov
2. Boyd (2), 8.
3. Boyd and Pyle, 228.
4. Alexandrov (2), 554. Alexandrov's emphasis.
5. Cavalli-Sforza, 65–66.
6. See Maller (3) and (4).
7. Punnett, 3.
8. Gould (2), 35–39.
9. Alexander, 20.
10. Alexander, 12. Alexander's emphasis.
11. Alexander, 6, note 18.
12. Appel (1), 340–41.
13. Mary S. Gardner, 717–18.
14. Ouspensky, 6.
15. Alexandrov (1), 227.
16. Ouspensky, 59.
17. Ouspensky, 60
18. Coleridge 11, i, 542.
19. Alexandrov (2), 550.
20. Mason, 50.

Appendix

James Joyce was acutely aware of number in all its vibrant significance, and his intellect was grounded in Catholic and medieval thought. So in *Finnegans Wake*, the total structure is built upon a universal interpretation of the four-fold:

> "Freely adapting Vico's three ages (theocratic, heroic, human) and his three institutions (religion, marriage and burial, and making a fourth the recurrence of the first, Joyce divides *Finnegans Wake* into 4 parts. The last words of the fourth, 'A way a love a last a loved a long the', begin a sentence which is completed by the opening paragraph of the first: 'riverrun, past Eve and Adam's, from swerve of shore to bend of bay, brings us by a commodius vicus of recirculation back to Howth Castle and Environs.' At intervals throughout the book all four parts are condensed into a series of four words or phrase; thunderburst, ravishment, dissolution and providentiality'; 'eggburst, eggblend, eggburial, and hatch-as-hatch-can'; 'sullemn fulminance, sollemn nuptualism, sallemn sepulture and providential divining.'"[1]

And:

> "As Homer's story, freely adapted, determines the three-part structure of *Ulysses* and the sequence of chapters, so Vico's system, freely adapted, determines the four-part structure of the *Wake* and the sequence of its chapters."[2]
>
> "Even the title, with which Joyce said he was making 'experiments', reflects the cyclic structure of the book: three syllables in a group are followed by a fourth, the 'Wake', just as the three long Books forming the cycle proper of *Finnegans Wake* are followed by the coda of Book IV, the Book of Waking. Similarly, the title may be read '*Fin negans Wake*', thus revealing possiibilities of cyclic endlessness."[3]

"There is one vital point of structure distinguishing the Viconian cycles from almost all the cyclic patterns which obsessed Blavatsky, for whereas Vico's theories are based on a tripartite formula, with a short interconnecting link between cycles, nearly every Indian system uses a primarily four-part cycle, with or without a short additional Fifth Age. Since in some respects a four-part cycle suits Joyce's purposes better than does a three-part, he extends Vico's Fourth Age, on the analogy of the Indian cycles, and gives it a great deal more detailed attention than it receives in *Scienza Nuova*."[4]

"Though the Fourth Age is sometimes described in Finnegans Wake as a disintegration—'O'c'stle, n'wc'stle, tr'c'stle, crumbling!'—it is no less frequently interpreted as a vital reorganisation of scattered forces, a resurrection which is positive rather than potential: 'hatch-as-hatch-can'. Earwicker mysteriously rises from the dead in I 4, departs, and is tried in absentia; Anna Livia babbles in lively fashion in I 8; Tristan and Iseult embrace in II 4; the new generation is carefully nurtured in the cradles of III 4. Joyce has extended the Fourth Age in each case by allowing it not only to prepare for, but to some extent also to recapitulate, in dream-like anticipation, the First Age of the cycle."[5]

Within the 3 Viconian Ages of Books I, II and III, Joyce creates four 4-chapter cycles. These 4 cycles are identified with the four Greco-medieval elements.

And there is the systole and diastole of 4 and 3. The two numbers merge and then part again, male and female:

"There are three children, but Isolde has a double, making a fourth; the four evangelists each have a house, but one of them is invisible since it is no more than a point in space. . . . The four, whose geometrical positions outline the frame which encloses the book, are themselves the 'fourdimmansions'—the synoptic gospellers corresponding to the three space dimensions, and Johnny, always late, to time. With Einstein and Minkowski at his back Joyce was able to surpass even the ancient mystics in complexity and tortuousness."[6]

With Flann O'Brien, numbers give the shape to everyday demotic life. But in *At Swim-Two-Birds*, number has become tyrannical, squeezing out humanly vital life. Person has dissolved, and a certain staleness and aridity hang over things. The novel begins and ends on 3. There is bread "for three minutes chewing" at the outset, from which there is retreat into "the privacy of my mind" amid three openings. And the conclusion will record:

"Numbers, however, will account for a great proportion of unbalanced and suffering humanity. One man will rove the streets seeking motor-cars with numbers that are divisible by seven. Well known, alas, is the case of the poor German who was very fond of three and who made each aspect of his life a thing of triads. He went home one evening and drank three cups of tea with three lumps of sugar in each, cut his jugular with a razor three times and scrawled with a dying hand on a picture of his wife good-bye, good-bye, good-bye."[7]

For O'Brien, 3 would seem to be especially connected with the matter of Ireland. Trellis's love of learning led him to "a three-volume work on the subject of the Irish monastic foundations at the time of the Invasion ," though he is sufficiently disturbed by the fact that "the three volumes by his bedside were blue," rather than the beloved green, to have them destroyed.[8] Finn MacCool is especially, though not exclusively, associated with the number 3, and this is embedded in his trilocular reflections:

"I incline to like the pig-grunting in Magh Eithne,/the bellowing of the stag of Ceara,/the whinging of fauns in Derrynish./The low warble of water-owls in Loch Barra also, sweeter than life that. I am fond of wing-beating in dark belfries,/cow-cries in pregnancy,/trout-spurt in a lake-top. . . . Soothing to my ear is the shout of a hidden black-bird,/the squeal of a troubled mare,/the complaining of wild-hogs caught in snow."[9]

There is a suggestion that encapsulating reality is focussed in the number 4, while the subjective dwells in the number 3. The "four walls of the house" envelop the triadic experiences and desires. The bet on "the 4.30 at Gatwick on Friday," is "4 to 1, 4/-" and has its counterpart in the "three stouts, called Kelly."[10]

Of all twentieth century attempts to locate the numerological significance of the cosmos, the greatest if most superstitious, is W.B. Yeats's *A Vision*. In one of Yeats's early drafts in the form of a dialogue, Michael Robartes defined Blake as conceiving "man as fourfold, while in his Mind, and as three-fold now that he is fallen, and I find that I must follow him." This would bear out Holub's observation that the present moment—at least in the fallen world—lasts three seconds. Yeats took his lead from Blake on this. As one critic puts it:

"His cosmic vision was essentially and consistently tetradic, based upon such occult sources as the Cabal, Neoplatonism, Boehme, and Blake. Besides 'The Table of Four Faculties,' Yeats discovered then other tetradic lists of characteristics in the human psyche, and numerous other important tetradic divisions are listed in this notebook: especially, Head, Heart, Loins and Fall as they are related to four zodiacal signs and four cardinal points, four Daimons, and four Memories ('declared to be frustration')."[11]

By the time of the 1925 version of *A Vision*, he had developed a hexadic conception of life that he compressed into the four Faculties; the Will, the Creative Mind, the Body of Fate and the Mask. "The *Will* and *Mask* are predominantly

Lunar or *antithetical*, the *Creative Mind* and the *Body of Fate* predominantly Solar or *primary*. When thought of in isolation, they take upon themselves the nature now of one phase, now of another."[12] Within these, there are:

"FOUR TYPES OF WISDOM
Wisdom of Desire
Wisdom of Intellect
Wisdom of Heart
Wisdom of Knowledge

FOUR CONTESTS
Moral
Emotional
Physical
Spiritual or supersensual

RAGE, PHANTASY, ETC.
Rage
Spiritual or supersensual Rage
Phantasy
Power"

There are four Perfections: Self-sacrifice, self-knowledge, unity of being, sanctity. And there is the quaternal psychological scheme in the Table of the Quarters as Yeats proceeds to explain "The Way of the Soul between the Sun and the Moon:"

"THE FOUR CONTESTS OF THE ANTITHETICAL WITHIN ITSELF

First quarter	With body.	
Second quarter	With heart.	In the first quarter body
Third quarter	With mind.	should win, in the second
Fourth quarter	With soul.	heart, and so on.

FOUR AUTOMATONISMS

First quarter	Instinctive.
Second quarter	Imitative.
Third quarter	Creative
Fourth quarter	Obedient

FOUR CONDITIONS OF THE WILL

First quarter	Instinctive.
Second quarter	Emotional.
Third quarter	Intellectual.
Fourth quarter	Moral.

FOUR CONDITIONS OF THE MASK

First quarter	Intensity (affecting Third Quarter).
Second quarter	Tolerance (affecting Fourth Quarter).
Third quarter	Convention of systematization (affecting First Quarter).
Fourth quarter	Self-analysis (affecting Second Quarter).

DEFECT OF FALSE CREATIVE MIND WHICH BRING THE FALSE MASK

First quarter	Sentimentality.
Second quarter	Brutality (desire for root facts of life).
Third quarter	Hatred.
Fourth quarter	Insensitiveness."[13]

However ultimately, for all its ingenious elaborateness, Yeats is providing a series of variations on the medieval cosmology. There is an intellectual ponderousness as he lists the Elemental Attributions of the four quarters in earth, water, air and fire. Joyce surpasses him by abstracting the dialectic of 4 and 3 and dramatizing and embodying it in his characters.

The longlastingness of the fourfold 'meme' well into the twentieth century is exemplified in Dylan Thomas. Thomas wrote that "when I experience anything I experience it as a thing and a word at the same time,"[14] and this intuitive faculty is crossed with an almost medieval world-picture. So in "Fern Hill:"

"Analysis of the means of plenitude discloses that Thomas has ordered his world in fours. Just as there are four kinds of plants (apples, grass, daisy, barley) and four of animals, all in the plural (calves, foxes, horses, men). There are also four kinds of birds in the plural (owls, nightjars, pheasants, swallows); though one singular bird, the false or sexual cock, intrudes on this harmonious scene of concordant fours. If this order is intentional, it probably refers to the creative *tetractys*, as the four elements in [section] *iii* certainly do."[15]

Elsewhere Thomas refers to counting "four elements and five senses, and man a spirit in love." And the cosmic 7 as 4+3 "may be expressed in the 4/3 distribution of creative 'greens' between the two halves of 'Fern Hill.'"[16] Realization of the self is often expressed in terms of the biblical worm, or larva, as potential transformation, as the source of all the striving in Thomas's poetry, perhaps also the legendary winged serpent, Eros. So it appears in "Before I Knocked" and {Alternise by Owl-light}, while in "Wintry Fever," the possibility of rebirth is glimpsed in "And I am dumb to tell the lover's tomb/How at my sheet goes the same crooked worm."

In T.S. Eliot's "Four Quartets,"

"the 'thematic material' of the poem is not an idea or a myth, but partly certain common symbols. The basic symbols are the four elements, taken as the material of moral life, and another way of describing *Four Quartets* . . . would be to say that 'Burnt Norton' is a poem about air, on which whispers are borne, intangible itself, but the medium of communication; 'East Coker' is a poem about earth, the dust of which we are made and into which we shall return; it tells of 'dung and death', and the sickness of the flesh; 'The Dry Salvages' is a poem about water which some Greek thinkers thought was the primitive material out of which the world arose, and which man has always thought of as surrounding and embracing the land, limiting the land and encroaching on it, itself illimitable; 'Little Gidding' is a poem about fire, the purest of the elements, by which some have thought the world would end, fire which consumes and purifies. We could say then that the whole poem is about the four elements whose mysterious union makes life, pointing out that in each of the separate poems all four are present; and perhaps adding that some have thought that there is a fifth element, unnamed but latent in all things: the quintessence, the true principle of life, and that this unnamed principle is the subject of the whole poem."[17]

However these patterns cross all cultural divides. The Oglala tribe is section of one of the Sioux, and part of the Lakota, from the region of the Missouri River on the plains of Dakota and Nebraska. Their belief and ritual centres on the fourfold:

"The category of the Gods as held by the shamans place them in four ranks with four in each rank, having prestige and precedence according to rank and place in rank.

The first rank is of the Superior Gods who are *Wi* (the Sun), the chief of the Gods; *Shan* (the Sky), the Great All-powerful Spirit; *Maka* (the Earth), the ancestress of all upon the world and provider for all; and *Inyan* (the Rock), the primal source of all things.

The second rank is of the Associate Gods who are *Hanwi* (the Moon), created by *Wi* to be his companion; *Tate* (the Wind0, created by *Skan* to be his companion; *Unk* (Contention), created by *Maka* to be her companion, but who was cast into the waters and is the Goddess of the Waters and ancestress of all evil beings; and *Wakinyan* (Winged One), created by *Inyan* to be his active associate.

The third rank is of the four Subordinate Gods who are *Ta Tanka* (The Buffalo God), the patron of ceremonies, of health, and of provision; *Hu Nonp* (the Bear God), the patron of wisdom; *Wani* (the Four Winds0, the vitalizer and weather; and *Yum* (the Whirlwind), the God of chance, of games, and of love.

The fourth rank is of the Inferior Gods who are *Nagi* (the Spirit); *Niya* (the Ghost); *Sicun* (the Intellect); and *Nagila* (the immaterial self of irrational things)."[18]

"A *wicasa wakan* (holy man or shaman) represents *Wakan Tanka* and speaks for him. . . . The shamans address *Wakan Tanka* as *Tobtob Kin*. This is in the speech that only the shamans should know . . . *Tobtob Kin* are Four-times-four Gods while *Tob-Kin* is only the Four Winds. The Four Winds is a God and is the *akicita* (messenger) of all the other Gods. The Four-times-four are *Wikan* [Sun] and *Hanwickan* [Moon]; *Taku Skanskan* [That which moves (Sky)] and *Tatekan* [Wind]; *Tob Kin* [The Four (Winds)] and *Yumnikan* [Whirlwind]; *Makakan* [Earth] and *Wohpe* [the Beautiful Woman]; *Inyankan* [Rock] and *Wakinyan* [Thunder Being]; *Tatankakan* [Buffalo Bull] and *Hunonpakan* [Two Legged (Grizzly Bear)]; *Wanagi* [Human Spirit] and *Woniya* [Human Life]; and *Nagila* [Nonhuman Spirit] and *Wasicunpi* [Guardian Spirits]. These are the names of the Good Gods as they are known to the people."[19]

It is the Scottish poet, Hugh MacDiarmid, who in his last major work defines the dynamic challenge. In his 1955 *In Memoriam James Joyce: A Vision of World Language*, he wrote:

> "There lie hidden in language elements that effectively combined
> Can utterly change the nature of man:
> Even as the recently-discovered plant growth hormone,
> Indole-acetic acid, makes holly cuttings in two months
> Develop roots that would normally take two years to grow,
> So perchance can we outgrow time
> And suddenly fulfil all history
> Established and to come."[20]

NOTES

1. Hart, 31.
2. Tindall, 10
3. Hart, 45–46.
4. Hart, 50.
5. Hart, 51–52.
6. Hart, 63–64
7. O'Brien, 217–18.
8. O'Brien, 100.
9. O'Brien, 14. The divisions are mine, only going to prove that MacCool speaks an incipient poetry.
10. O'Brien, 37–38.
11. Yeats (1), xlix, xxvii-xxviii.
12. Yeats (1), 14.
13. Yeats (2), 144–46.
14. Reid, 20.
15. Walford Davies, 248.
16. Walford Davies, 248.
17. Helen Gardner, 126.
18. Walker, 50–51
19. Walker, 94.
20. MacDiarmid, 55.

Bibliography

Abbott, J. "Music, maestro, please!" *Nature* 416 (March 7, 2002): 12–14.

Abrahams, Cecil Anthony. *William Blake's Fourfold Man.* Bonn: Bouvier, 1978.

Ackroyd, Peter. *Blake.* London: Sinclair-Stevenson, 1995.

Albert, David Z. *Quantum Mechanics and Experience.* Cambridge, Mass.: Harvard University Press, 1992.

Alexander, Victoria N. "Neutral Evolution and Aesthetics: Vladimir Nabokov and Insect Mimicry." Santa Fe: Santa Fe Institute Working Papers, no. 01-10-057, 2001.

Alexandrov, Vladimir E.(1). *Nabokov's Otherworld.* Princeton: Princeton University Press, 1991.

—— (2), ed. *The Garland Companion to Vladimir Nabokov.* London: Garland, 1995.

Alford, John A. ed. *A Companion to Piers Plowman.* Berkeley and Los Angeles: University of California Press, 1988.

Allen, Elizabeth Cheresh and Gary Saul Morson, eds. *Freedom and Responsibility in Russian Literature: Essays in Honor of Robert Louis Jackson.* Evanston: Northwestern University Press, 1995.

Alonso, Dámaso (1). *Ensayos sobre Poesía Española.* Buenos Aires: Revista de Occidente, 1946.

—— (2). *Gozos de Vista y otros poemas.* Madrid: Espasa-Calpe, 1981.

—— (3). *Hijos de la ira: Diario Intimo.* ed. Miguel J. Flys. Madrid: Castalia, 1986.

—— (4). *Obras Completas 1. Estudios Lingüísticos Peninsulares.* Madrid: Gredos, 1972.

—— (5). *Poemas Escogidos.* Madrid: Gredos, 1969.

—— (6). *Poetas Españoles Contemporáneos* . Madrid: Gredos, 1965.

Altenmüller, Eckhart O. "Music in your Head." *Scientific American.* New York: Scientific American Special on Mind (2004): 24–31.

Amis, Robin. *A different Christianity: Early Christian Esotericism and Modern Thought.* Albany: State University of New York, 1995.

Anonymous. *Sir Gawayne and the Grene Knight.* Pp. 349–428 in Ford.

157

Apollinaire, Guillaume. *Méditations esthétiques—Les peintres cubistes.* edited by L.C. Breunig and J.-Cl. Chevalier. Paris: Miroirs de l'Art, 1965.

Appel, Alfred, Jr. (1). *The Annotated Lolita.* New York: McGraw-Hill, 1991.

—— (2)."*Lolita*: the Springboard of Parody." Pp. 106–203 in Dembo.

—— (3) and Charles Newman, eds. *Nabokov: criticism, reminiscences, translations and tributes.* London: Weidenfeld & Nicolson, 1971.

Apuleius (1). *The Golden Asse of Apuleius.* trans. William Adlington. London: Grant Richards, 1913.

—— (2). *Metamorphoses.* ed. J. Arthur Hanson. Cambridge, Mass.: Harvard University Press, 1989.

—— (3). *The Works of Apuleius.* London: Bell, 1889.

Aristotle (1). *The Athenian Constitution: the Eumedian Ethics: On Virtues and Vices.* trans. Harris Rackham. Cambridge, Mass.: Harvard University Press, 1935.

—— (2), *De Anima.* ed. David Ross. Oxford: Clarendon, 1961.

—— (3), *Metaphysics.* ed. John Warrington. New York: Dutton, 1956.

Arnold, Armin, ed. *The Symbolic Meaning.* Fontwell: Centaur, 1962.

Arnold, Matthew. *The Complete Prose Works* 9. ed. R.H. Super. Ann Arbor: University of Michigan Press, 1973.

Arvin, Newton. *Herman Melville.* London: Methuen,1950.

Atwell, John E. *Schopenhauer on the Character of the World: the Metaphysics of Will.* Berkeley and Los Angeles: University of California Press, 1995.

Augustine (1). *The City of God against the Pagans.* ed. R.W. Dyson Cambridge: Cambridge University Press, 1998.

—— (2). *De Libero Arbitrio.* Indianopolis: Hackett, 1993.

Ausmus, Harry J. *A Schopenhauerian Critique of Nietzsche's Thought: Toward a Restoration of Metaphysics.* Lewiston: Mellen Press, 1996.

Balashov, Yuri (1). "Enduring and Perduring Objects in Minkowski Space-Time." *Philosophical Studies*, 99: 129–66.

—— (2). "Relativistic Objects." *Noûs*, 33 (1999): 644–62.

Baldick, Julian. "Patterns of prophecy." *Times Literary Supplement* (8 August 2003): 26.

Ballantine, Christopher. *Twentieth Century Symphony.* London: Dobson, 1983.

Barnett, F.J. et.al. (eds). *History and Structure of French: Essays in Honour of Professor T.B.W. Reid.* Oxford: Blackwell, 1972.

Barrow, John D.(1). *The Artful Universe.* Oxford: Clarendon, 1995.

—— (2). *The Constants of Nature: from Alpha to Omega.* London: Cape, 2002.

—— (3) and Frank J. Tipler. *The Anthropic Cosmological Principle.* Oxford: Clarendon, 1986.

Bate, Walter Jackson. *Coleridge.* London: Weidenfeld & Nicolson, 1969.

Bates, Henry Walter (1). *Naturalist on the River Amazons.* London: Dent, 1969.

—— (2). "Contributions to an insect fauna of the Amazon valley. Lepidoptera: Heliconidae." *Transactions of the Linnean Society of London*, 23 (1862): 495–566.

Battersby, James L. *Paradigm Regained: pluralism and the practice of criticism.* Philadelphia: University of Pennsylvania Press, 1991.

Batts, Michael S.(1). "Numbers and Number Symbolism in Medieval German Poetry," *MLQ*, 24 (1963): 342–49.

—— (2). "The Origins of Numerical Symbolism and Numerical Patterns in Medieval German Literature," *Traditio* XX (1964): 462–71.

Bayley, Harold. *The Lost Language of Symbolism: an enquiry into the origin of certain letters, words, names, fairy-tales, folklore, and mythologies*. New York; Barnes & Noble, 1968.

Beckwith, Isbon T. *The Apocalypse of John: Studies in Introduction*. Michigan: Baker Book House, 1967.

Beddall, Barbara Gould, ed. *Wallace and Bates in the Tropics: an introduction to the theory of natural selection*. London: Collier-Macmillan, 1969.

Bede, Venerable. *De Temporum Ratione* in *Bedae Opera de temporibus*, ed. Charles W. Jones. Cambridge, Mass.: Medieval Academy, 1943.

Bely, A. *Selected Essays of Bely*. ed., Steven Cassidy. Berkeley and Los Angeles: University of California, 1985.

Bergonzi, Bernard. *T.S. Eliot: Four Quartets*. London: Macmillan, 1979.

Berlin, Brent and Paul Kay. *Basic Color Terms: their universality and evolution*. Berkeley and Los Angeles: University of California Press, 1969.

Berlioz, Hector. *A Critical Study of Beethoven's Nine Symphonies*. London: Reeves, 1958.

Berry, Andrew, ed. *Infinite Tropics: An Alfred Russel Wallace Anthology*. London. New York: Verso, 2002.

Berry, Francis, "*Sir Gawayne and the Grene Knight*." Ford: 146–56.

Bertulani, C.A. and V. Zelevinsky, "Is the tetraneutron a bound dineutron-dineutron molecule?" *Journal of Physics G: Nuclear and Particle Physics* 29 (October 2003): 2431 37.

Bezanson, Walter E. "*Moby-Dick*: Work of Art." Pp. 651–71 in *Moby-Dick*, edited by Harrison Hayford and Hershel Parker. New York: Norton, 1967.

Bickel, Ernst. "Homerischer Seelenglaube," *Schriften der Königsberger Gelehrten Gesellschaft* (1925).

Bickerton, Derek. *Language & Species*. Chicago: University of Chicago Press, 1990.

Blackham, H.J. *The Fable as Literature*. London: Athlone, 1985.

Blackmore, Susan. *The Meme Machine*. Oxford: Oxford University Press, 2000.

Blackwell, Stephen, "The Poetics of Science in, and around, Nabokov's *The Gift*," *The Russian Review*, 62, no. 2 (April 2003): 243–61.

Blake, William (1). *The Letters*. ed., Geoffrey Keynes. London: Hart-Davis, 1956.

—— (2). *The Poems*. ed., W.H. Stevenson. New York: Norton, 1972.

Bloch, Ernst. *Essays on the Philosophy of Music*. trans. Peter Palmer. Cambridge: Cambridge University Press, 1985.

Boardman, Philip. *The World of Patrick Geddes*. London: Routledge & Kegan Paul, 1978.

Boberg, I.M., "The Tale of Cupid and Psyche," *Classica et Mediaevalia* 1 (1938): 177–216.

Boden, Margaret A. *The Philosophy of Artificial Intelligence*. Oxford: Oxford University Press, 1990.

Böhme, Joachim. *Die Seele und das Ich im homerischen Epos. Mit einem Anhang: Vergleich mit dem Glauben der Primitiven.* Leipzig and Berlin: B.G. Teubner, 1929.

Boisacq, Emile. *Dictionnaire Etymologique de la Langue Grecque.* Heidelberg: Winter, 1950.

Boston, Thomas. *Human Nature in its Fourfold State.* Edinburgh: Hamilton, Balfour & Neill, 1752.

Bové, Paul A. *Deconstructive Poetics: Heidegger and Modern American Poetry.* New York: Columbia University Press, 1980.

Bowen, Meirion (1), ed. *Music of the Angels: Essays and Sketchbooks of Michael Tippett.* London: Eulenburg. 1980.

—— (2), ed. *Tippett on Music.* Oxford: Clarendon, 1995.

Bowersock, G.W. *Hellenism in Late Antiquity.* Cambridge: Cambridge University Press, 1990.

Bowler, Peter J.(1). *Charles Darwin: the years of controversy: the Origin of Species and its critics 1859–1882.* London, 1972.

—— (2). *The Non-Darwinian Revolution: Reinterpreting a Historical Myth.* Baltimore: Johns Hopkins University Press, 1988.

Boyd, Brian (1), "Literature and Discovery," *Philosophy and Literature*, 23:2 (October 1999): 312–33.

Boyd, Brian (2) and Robert Michael Pyle, eds. *Nabokov's Butterflies: unpublished and uncollected writings.* Harmondsworth: Allen Lane, 2000.

Brandon, S.G.F. *History, Time and Deity.* Manchester: Manchester University Press, 1965.

Bremmer, Jan. *The Early Greek Concept of the Soul.* Princeton: Princeton University Press, 1983.

Brenner, S., J.D. Murray, L. Wolpert, eds. *Theories of Biological Pattern Formation: A Royal Society Discussion.* London: Royal Society, 1981.

Briggs, Katherine M. *A Dictionary of British Folk-Tales.* London: Routledge & Kegan Paul,1970.

Brod, Max, ed. *The Diaries of Franz Kafka.* London: Secker & Warburg, 1948.

Brodhead, Richard H., ed. *New Essays on Moby-Dick.* Cambridge: Cambridge University Press, 1986.

Bronowski, Jacob (1). *The Origins of Knowledge and Imagination.* New Haven: Yale University Press, 1978.

—— (2). *Science and Human Values.* London: Hutchinson, 1961.

Brooks, John Langdon. *Just before the Origin: Alfred Russel Wallace's theory of evolution.* New York: Columbia University Press, 1984.

Brotherston, Gordon. *Book of the Fourth World.* Cambridge: Cambridge University Press, 1992.

Brown, Norman O. *Love's Body.* New York: Vintage, 1966.

Buckland, A.W., "Four as a Sacred Number," *The Journal of the Anthropological Society of Institute of Great Britain and Ireland*, XXV (1896): 96–102.

Buell, Lawrence. *The Environmental Imagination: Thoreau, Nature Writing and the Formation of American Culture.* Cambridge, Mass.: Belknap Press, 1966.

Burkhardt, Frederick and Sydney Smith. eds. (1). *The Correspondence of Charles Darwin* 5: 1851–1855. Cambridge: Cambridge University Press, 1989.

—— (2), *The Correspondence of Charles Darwin* 6:1856–1857. Cambridge: Cambridge University Press, 1990.

Burnet, J., "The Socratic Doctrine of the Soul," *Proceedings of the British Academy* (1916): 235–59.

Burnett, James (Lord Monboddo). *Of the Origin and Progress of Language*. Edinburgh: Kincaid & Creech, 1773–1792.

Burrow, J.A. *The Ages of Man: A Study in Medieval Writing and Thought*. Oxford: Clarendon, 1988.

Butler, Christopher (1). *Number Symbolism*. London: Routledge & Kegan Paul, 1970.

—— (2). "Numerological Thought." Pp. 1–31 in Fowler (2).

Butler, Samuel (1). *Evolution, Old and New; or, The Theories of Buffon, Dr. Erasmus Darwin, and Lamarck, as compared with that of Mr. Charles Darwin*. London: Hardwicke & Bogue, 1879.

—— (2). *Luck, or Cunning, as the main means of organic modification?* London: A.C. Fifield, 1920.

—— (3). *Unconscious Memory*. London: A. C. Fifield, 1920.

Butterworth, Brian. *The Mathematical Brain*. London: Macmillan, 1999.

Byrhtferth's Manual A.D. 1011. ed., S.J. Crawford. Oxford: Oxford University Press, 1929.

Calvin, William H. and Derek Bickerton. *Reconciling Darwin and Chomsky with the Human Brain*. Cambridge: A Bradford Book, MIT Press, 2000.

Camerini, Jane R., ed. *The Alfred Russel Wallace Reader*. Baltimore: Johns Hopkins, 2002.

Candler, Howard, "On the Symbolic Use of Number in the 'Divina Commedia' and Elsewhere," *Transactions of the Royal Society of Literature* 30, no. 2 (1910): 1–29.

Carr, B.J., and M.J. Rees, "The anthropic principle and the structure of the physical world," *Nature* 278 (1979): 605–12.

Caspar, Max, *Kepler*. trans. C. Doris Hellman. London: Abelard-Schuman, 1959.

Cavalli-Sforza, Luigi., and Marcus Feldman. *Cultural Transmission and Evolution: A Quantitative Approach*. Princeton: Princeton University Press, 1981.

Chambers, Robert. *Vestiges of the Natural History of Creation*. Edinburgh: Churchill, 1884.

Chase, Richard, ed. *Melville: A Collection of Critical Essays*. Englewood Cliffs: Prentice-Hall, 1962.

Chénier, André, ed. Gérard Walter. *Oeuvres complètes*. Paris: Pléiade, 1950.

Chomsky, Noam. *New Horizons in the Study of Language and Mind*. Cambridge: Cambridge University Press, 2000.

Chown, Marcus (1). *The Universe Next Door: Twelve mind-blowing ideas from the cutting edge of science*. London: Headline, 2001.

—— (2). "It all fits . . ." *New Scientist* (19 June, 1999): 18.

—— (3). "Why should nature have a favourite number?" *New Scientist* (21/28 December 2002): 55–56.

Citati, Pietro. *Kafka*. London: Minerva, 1991.

Claus, David B. *Toward the Soul: an inquiry into the meaning of ψυχή before Plato*. New Haven: Yale University Press, 1981.

Clodd, Edward. *Pioneers of Evolution*. London: Cassell, 1907.

Coburn, Kathleen (1). *Inquiring Spirit*. London: Routledge & Kegan Paul, 1951.

—— (2) and Merton Christiansen, eds. *The Notebooks of Samuel Taylor Coleridge*. London: Routledge & Kegan Paul, 1957–90.

Cohen, Michael P. *The Pathless Way: Muir and American Wilderness*. Madison: Wisconsin University Press, 1984.

Coleridge, S.T. (1). *Collected Works*. Princeton: Bollingen Series, Princeton University Press, 1980–2001.

—— (2). *The Letters of S.T. Coleridge*. ed., E.H. Coleridge. London: Heinemann, 1895.

Conrad, Joseph (1). *Almayer's Folly: A Story of an Eastern River*. London: Dent, 1961.

—— (2). *Lord Jim*. London: Dent, 1923.

Cornford, F.M.(1). *From Religion to Philosophy: A Study of the Origins of Western Speculation*. London: Arnold, 1912.

—— (2). *Greek Religious Thought: from Homer to the Age of Alexander*. New York: Dutton, 1923.

Cronin, Helena. *The Ant and the Peacock: altruism and sexual selection from Darwin to today*. Cambridge: Cambridge University Press, 1991.

Curtius, Ernst Robert. *European Literature and the Latin Middle Ages*. London: Routledge & Kegan Paul, 1953.

Cysarz, D; Von Bonin, D; Lackner, H; Moser M; Bettermann, H."Oscillations of heart rate and respiration synchronize during poetry recitation," *American Journal of Physiology—Heart and Circulatory Physiology*, 287 no. 2. (September 2004): H579.

Damon, S. Foster. *A Blake Dictionary*. Rhode Island: Brown University Press, 1965.

Dante, *La vita nuova*. trans. Barbara Reynolds. Harmondsworth: Penguin, 1969.

Darcus, S.M. "A Person's Relation to ψυχή in Homer, Hesiod, and the Greek Lyric Poets," *Glotta* 57 (1979): 30–39.

Darwin, Charles (1). *The Descent of Man, and selection in relation to sex*. Princeton: Princeton University Press, 1981.

—— (2). *The Origin of Species*. ed. Gillian Beer. Oxford: Oxford University Press, 1998.

—— (3), and T.H. Huxley. *Autobiographies*. Oxford: Oxford University Press, 1974.

—— (4), and A.R. Wallace. *Evolution by Natural Selection*. Cambridge: Cambridge University Press, 1958.

Davie, George. *The Scotch Metaphysics: A century of Enlightenment in Scotland*. New York: Routledge, 2001.

Davies, Paul (1). *Other Worlds*. Harmondsworth: Penguin, 1990.

—— (2). *The Search for a Grand Unified Theory of Nature*. London: Heinemann, 1984.

—— (3), ed. *The New Physics*. Cambridge: Cambridge University Press, 1989.

Davies, Walford *Dylan Thomas: New Critical Essays*. London: Dent, 1972.

Dawkins, Richard (1). *The Blind Watchmaker*. London: Longman, 1986.

—— (2). *The Extended Phenotype: the Gene as the Unit of Selection*. Oxford: Freeman, 1982.

—— (3). *River out of Eden: A Darwinian View of Life*. London: Weidenfeld & Nicolson, 1995.

—— (4). *The Selfish Gene*. Oxford: Oxford University Press, 1989.

—— (5), "In Defence of Selfish Genes," *Philosophy*, 56 (1981): 556–73.

De Ricord, Elsie Alvarado. *La Obra Poética de Dámaso Alonso*. Madrid: Gredos, 1968.

De Troyes, Chrestien. *Yvain (Le Chevalier au lion)*. ed. T.B.W. Reid Manchester: Manchester University Press, 1942.

Deacon, Terrence. *The Symbolic Species: the co-evolution of language and the human brain*. New York: Norton, 1997.

Debicki, Andrew P. *Dámaso Alonso*. New York: Twayne, 1970.

Defries, Amelia. *The Interpreter: Geddes—the man and his gospel*. London: Routledge, 1927.

Dembo, L. S. (ed.). *Nabokov: the man and his work*. Madison: University of Wisconsin Press, 1967.

Dennett, Daniel C. (1). *Consciousness Explained*. Harmondsworth: Allen Lane, 1991.

—— (2). *Darwin's Dangerous Idea: evolution and the meanings of life*. Harmondsworth: Allen Lane, 1995.

—— (3). "Memes and the Exploitation of Imagination," *Journal of Aesthetics and Art Criticism*, 48 no. 2 (Spring 1990): 127–35.

Descartes, René (1). *Cogitationes privatae*. ed. Charles Adam and Paul Tannery, from *Oeuvres* X (Paris:Vrin, 1966): 213–48.

—— (2). *Rules for the Direction of the Mind*, trans. L.J. Lafleur. Indianapolis: Bobbs-Merrill, 1961.

Deutsch, Glenn. "Talent as Persistence: a profile of Bob Hicok," *Poets & Writers*, 32, no. 2 (March/April 2004): 40–45.

Devlin, Keith. "A matter of fractal," *Guardian* (April 17, 1997).

Dicke, R.H. "Dirac's Cosmology and Mach's Principle," *Nature*, 192 (November 4, 1961): 440–41.

Dirac, Paul A.M. (1), *Directions in Physics*. New York: Wiley, 1978.

—— (2). "A new basis for cosmology," *Proceedings of the Royal Society of London A*, 165 (1938): 199–208.

—— (3). "Reply to R.H. Dicke, *Nature* 192 (November 4, 1961): 441.

Donaldson, S.K. and P.B. Kronheimer. *The Geometry of Four-Manifolds*. Oxford: Clarendon, 1990.

Duckworth, George E. "Mathematical Symmetry in Vergil's *Aeneid*," *Transactions and Proceedings of the American Philological Association*, XCI (1960): 184–220.

Dumézil, Georges. *The Destiny of the Warrior*. Chicago: University of Chicago Press, 1969.

Durant, John. "Scientific Naturalism and Social Reform in the Thought of Alfred Russel Wallace," *British Journal for the History of Science* 12: pt. 1. no. 40: 31–58.

Eagleton, Terry (1). *The Ideology of the Aesthetic*. Oxford: Blackwell, 1990.

—— (2). "Bakhtin, Schopenhauer, Kundera." Pp. 229–40 in Hirschkop.

Eccles, John C. *Evolution of the Brain: creation of the self.* London: Routledge, 1991.

Edey, Maitland A. and Donald C. Johanson. *Blueprints: solving the mystery of evolution.* Oxford: Oxford University Press, 1990.

Edwards, Mark W. *Sound, Sense and Rhythm: listening to Greek and Latin Poetry.* Princeton: Princeton University Press, 2002.

Einarsson, Stefán. *A History of Icelandic Literature.* Baltimore: Johns Hopkins, 1957.

Einstein, Albert (1). *Relativity: the Special and General Theory.* New York: Crown, 1961).

—— (2). "Autobiographical Notes." Pp. 3–95 in Schlipp.

—— (3) and Leopold Infeld. *The Evolution of Physics: the growth of ideas from the early concepts to Relativity and Quanta.* Cambridge: Cambridge University Press, 1938.

Eisley, Loren (1). *Darwin's Century.* London: Gollancz, 1959.

—— (2). *The Immense Journey.* New York: Vintage, 1958.

—— (3). "Alfred Russel Wallace," *Scientific American* 200:2 (February 1959): 70–83.

Eliot, George. *Middlemarch.* Harmondsworth: Penguin, 1972.

Eliot, T.S.(1). *Four Quartets.* London: Faber, 1944.

—— (2). *Selected Essays.* London: Faber, 1972.

Emerson, Ralph Waldo. *The Complete Works* 12. ed. Edward Waldo Emerson. Boston: Houghton Mifflin, 1903–04.

es-Said, Iasam and Ayse Parmun. *Geometric concepts in Islamic Art.* London: World of Islam Festival Publishing Co., 1976.

Evreinoff, Nicolas. *The Theatre in Life.* trans. Alexander I. Nazaroff. London: Harrap, 1927.

Farbridge, Maurice H. *Studies in Biblical and Semitic Symbolism.* New York: Dutton, 1923.

Farmelo, Graham. "Physics+Dirac=poetry." *Guardian* (February 21, 2002).

Farrer, Austin. *The Revelation of St. John the Divine.* Oxford: Oxford University Press, 1964.

Faulkner, William. *Intruder in the Dust.* London: Chatto & Windus, 1951.

Ferrera, Lawrence. "Schopenhauer on music as the embodiment of Will." Pp. 183–199 in Jacquette,.

Fichman, Martin. *An Elusive Victorian: the evolution of Alfred Russel Wallace.* Chicago: University of Chicago Press, 2004.

Fiedler, Leslie A. *Love and Death in the American Novel.* New York: Criterion, 1960.

Field, Andrew. *The Life and Art of Vladimir Nabokov.* London: Queen Anne Press, 1987.

Fisher, P.F. *The Valley of Vision: Blake as Prophet and Revolutionary.* Toronto: University of Toronto Press, 1961.

Ford, Boris. *The Age of Chaucer.* Harmondsworth: Penguin, 1974.

Fowler, Alastair (1). *Spenser and the Numbers of Time.* London: Routledge & Kegan Paul, 1964.

—— (2), ed. *Silent Poetry: Essays in Numerological Analysis.* London: Routledge & Kegan Paul, 1970.

Fowler, Douglas. *Reading Nabokov*. Ithaca: Cornell University Press, 1974.

Francis, Elizabeth A., ed. *Studies in Medieval French: presented to Alfred Ewart in honour of his seventieth birthday*. Oxford: Clarendon, 1961.

Francis, Richard Alan, "Foreword for the Forewarned," *Review of Contemporary Fiction*, 10 no. 3 (Fall 1990): 180–81.

Franklin, Peter. *The Idea of Music: Schoenberg and others*. London: Macmillan, 1985.

Franks, Daniel W. and Jason Noble, "Batesian mimics influence mimicry ring evolution," *Proceedings of the Royal Society of London* B 271 (2004): 191–96.

Friedberg. *An Adventurer's Guide to Number Theory*. New York: Dover, 1994.

Frost, Robert. *Complete Poems*. New York: Holt, Rinehart & Winston, 1959.

Frye, Northrop (1). *Fearful Symmetry: a study of William Blake*. Princeton: Princeton University Press, 1969.

—— (2). "Notes for a Commentary on *Milton*." Pp. 99–137 in Pinto (1).

Fuentes, Carlos. *Myself with Others*. London: Deutsch, 1988.

Gardner, Helen. "The Music of *Four Quartets*.". Pp. 119–37 in Bergonzi.

Gardner, Mary S. *The Biology of Invertebrates*. New York: McGraw-Hill, 1972.

Gardner, W.H. and N.H. McKenzie. *The Poems of Gerard Manley Hopkins*. Oxford: Oxford University Press, 1970.

Gaskill, P.H. "Hermann Broch as a Translator of Edwin Muir." *New German Studies*, 6 no. 2 (Summer 1978): 101–15.

Gaston, Kevin J. *Biodiversity: A Biology of Numbers and Difference*. Oxford: Blackwell, 1996.

Gaukroger, Stephen. *Descartes: an intellectual biography*. Oxford: Clarendon, 1995.

Gautier, Marie-Lise Gazarian. "An Interview with Julián Ríos." *Review of Contemporary Fiction* 10 no. 3 (Fall 1990): 182–88.

George, Wilma. *Biologist Philosopher: a study of the life and writings of Alfred Russel Wallace*. (London: Abelard-Schuman, 1964).

Gilbert, Lawrence I, Jamshed R. Tata, Burr G. Atkinson. *Metamorphosis: postembryonic reprogramming of Gene Expression in Amphibian and Insect Cells*. San Diego, Cal.: Academic Press, 1996.

Goehr, Lydia. "Schopenhauer and the musicians: an inquiry into the sounds of silence and the limits of philosophizing about music." Pp. 200–28 in Jacquette.

Gold, Barri J. "The Consolation of Physics: Tennyson's Thermodynamic Solution." *PMLA*, 117 no. 3 (May 2002): 449–464.

Gompf, Robert E. and András I. Stipsicz. *4–Manifolds and Kirby Calculus*. Providence, R.I.: American Mathematical Society, 1999.

Gonzalez, Guillermo (1). "Is the Sun anomalous." *Astronomy and Geophysics*, 40 no. 5 (October 1999): 25–29.

—— (2). "Wonderful Eclipses." *Astronomy and Geophysics*, 40 no. 3 (June 1999), 18–20.

Gould, Stephen Jay (1). *Bully for Brontosaurus: Further Reflections in Natural History*. Harmondsworth: Penguin, 1992.

—— (2). *Ever Since Darwin: reflections in natural history*. Harmondsworth: Penguin, 1978.

—— (3). *I Have Landed: splashes and reflections in Natural History*. London: Cape, 2002.

—— (4). *The Panda's Thumb: New Reflections in Natural History*. Harmondsworth: Penguin, 1990.

—— (5), "Darwinian Fundamentalism," *New York Review of Books* 44 no. 10 (June 12, 1997): 34–37.

Graves, Robert. *The Greek Myths*. Harmondsworth: Penguin, 1962.

Grayson, Jane (1), Arnold McMillin and Priscilla Meyer. *Nabokov's World* 1. Basingstoke: Palgrave, 2002.

—— (2). *The Shape of Nabokov's World* 2. Basingstoke: Palgrave, 2002.

Graz, Louis. "L'Iliade et la personne." *Esprit* 28 (1960): 1390–1403.

Greene, Brian. *The Elegant Universe*. London: Cape, 1999.

Grene, Marjorie, *Descartes* (Indianapolis: Hackett, 1998)

Guetti, James. *The Limits of Metaphor: a Study of Melville, Conrad and Faulkner*. Ithaca: Cornell University Press, 1967.

Guthrie, W.K.C. *A History of Greek Philosophy* 1. Cambridge: Cambridge University Press, 1962.

Hanby, Michael. *Augustine and Modernity*. New York: Routledge, 2003.

Harman, Alec, Wilfrid Mellers and Anthony Milner. *Man & His Music: the story of musical experience in the West*. (London: Barrie & Jenkins, 1973).

Harrison, S.J. *Apuleius: A Latin Sophist*. Oxford: Oxford University Press, 2000.

Hart, Clive. *Structure and Motif in Finnegans Wake*. London: Faber, 1962.

Hartmann, Franz (1). *The Life and Doctrines of Jacob Boehme: the God-taught philosopher*. London: Kegan Paul, Trench, Trübner, 1891.

—— (2). *The Life of Paracelsus*. London: Kegan Paul, Trench, Trübner, 1896.

Hatto, A.T. and R.J. Taylor. "Recent Work on the Arithmetical Principle in Medieval Poetry." *Modern Language Review* 46 (1951): 396–403.

Havelock, Eric A. (1). *The Literate Revolution in Greece and its cultural consequences*. Princeton: Princeton University Press, 1982.

—— (2). *The Muse Learns to Write: reflections on orality and literacy from antiquity to the present*. New Haven: Yale University Press, 1986.

Hegel, G.W.F. *The Philosophy of Fine Art* 3. trans. F.P.B. Osmaston. London: G. Bell, 1920.

Heidegger, Martin (1). *On the Way to Language*. trans. Peter D. Hertz. New York: Harper & Row, 1971.

—— (2). *Poetry, Language, Thought*. trans. Albert Hofstadter. New York: Harper & Row, 1975.

—— (3). *The Question Concerning Technology and Other Essays* trans. William Lovitt. New York: Harper Colophon, 1977.

Heller, Mark. *The ontology of physical objects: Four-dimensional hunks of matter*. Cambridge: Cambridge University Press, 1990.

Henderson, Linda Dalrymple. *The Fourth Dimension and Non-Euclidean Geometry in Modern Art*. Princeton: Princeton University Press, 1983.

Heninger, S.K., Jr. (1). *Touches of Sweet Harmony: Pythagorean Cosmology and Renaissance Poetics*. San Marino, Cal.: Huntingdon Library, 1974.

—— (2). "Some Renaissance Versions of the Pythagorean Tetrad." *Studies in the Renaissance*, 8 (1961): 7–35.

Hieatt, A. Kent. *Short Time's Endless Monument: the symbolism of the numbers in Edmund Spenser's Epithalamion.* Port Washington: Kennikat, 1972.

Hilton, Nelson. *Essential Articles for the study of William Blake.* Hamden, Conn.: Archon Books, 1986.

Hinton, C.H. (1). *The Fourth Dimension.* London: Swan Sonnenschein, 1904.

—— (2) *A New Era of Thought.* London: Swan Sonnenschein, 1988.

Hirschkop, Ken and David Shepherd. *Bakhtin and Cultural Theory.* Manchester: Manchester University Press, 2001.

Hofstadter, Douglas R. *Gödel, Escher, Bach: An Eternal Golden Thread.* New York: Basic Books, 1973.

Holub, Miroslav. *The Dimension of the Present Moment and other essays.* London: Faber, 1990.

Hopkins, Gerard Manley. *Poems.* ed. W.H. Gardner and N.H. MacKenzie. Oxford: Oxford University Press, 1970.

Hopper, Vincent Foster. *Medieval Number Symbolism: Its Sources, Meaning, and Influence on Thought and Expression.* New York: Columbia University Press, 1938.

Houston, Amy "Conrad and Alfred Russel Wallace." Pp. 29–48 in Knowles.

Hugo, Victor. *Les Contemplations.* Paris: Gallimard, 1973.

Hutton, James. "Theory of the Earth." *Transactions of the Royal Society of Edinburgh*, 1. (1788).

Immisch, Otto. "Sprachliches zum Seelenschmetterling." *Glotta* 6 (1915): 193–205.

Irvine, William. *Apes, Angels and Victorians.* Lanham, Md.: University Press of America, 1983.

Jacquette, Dale, ed. *Schopenhaeur, philosophy, and the arts.* Cambridge: Cambridge University Press, 1996.

James, Henry. *The Critical Muse: Selected Criticism.* ed. Roger Gard. Harmondsworth: Penguin, 1987.

Jastrow, Morris, Jr. *The Civilization of Babylonia and Assyria: its remains, language, history, religion, commerce, law, art, and literature.* Philadelphia: J.B. Lippincott, 1915).

Johnson, Kurt. "nabokov's fyodor takes wing." *Zembla.* 2001. <http://www.libraries.psu.edu/nabokov/fyodor.htm.> (18 Jan. 2005).

Johnson, Michael L. *Mind, Language, Machine: Artificial Intelligence in the Postmodernist Age.* Basingstoke: Macmillan, 1988.

Jones, Frederick L., ed. *The Letters of Shelley* 1. Oxford: Oxford University Press, 1964.

Jones, Steve. *Almost Like a Whale: the origin of species updated.* London: Black Swan, 2001.

Joron, Mathleu and James L. B. Mallet. "Diversity in mimicry: paradox or paradigm?" *TREE*, 13 no. 11 (November 1998): 462–66.

Joyce, James. *Finnegans Wake.* London: Faber, 1975.

Jung, Carl Gustav (1). *Collected Works* 6. ed. Herbert Read, Michael Fordham and Gerhard Adler. London: Routledge & Kegan Paul, 1964.

—— (2). *Psychology and Religion*. New Haven: Yale University Press, 1966.

Kalnins, Mara, ed. *D.H. Lawrence: Centenary Essays*. Bristol: Bristol Classical, 1986.

Kafka, Franz (1). *The Complete Short Stories*. ed. Nahum N. Glatzer. London: Minerva, 1992.

—— (2). *Dearest Father: Stories and Other Writings*. New York: Schocken Books, 1954.

—— (3). *Die Verwandlung*. ed. Marjorie L. Hoover. London: Methuen, 1962.

Kant, Immanuel (1). *Critique of the Power of Judgment*. ed. Paul Guyer. Cambridge: Cambridge University Press, 2000.

—— (2). *Critique of Pure Reason*. ed. Paul Guyer and Allen W. Wood. Cambridge: Cambridge University Press, 1998.

—— (3). *Critique of Practical Reason*. trans. Lewis White Beck. Indianapolis: Bobbs-Merrill Co., 1956 .

—— (4). *The Metaphysics of Morals*. ed. Mary Gregor. Cambridge: Cambridge University Press, 1996.

—— (5). *Prologomena to any Future Metaphysics that will be able to present itself As A Science*. trans. Peter G. Lucas. Manchester: Manchester University Press, 1966.

Karlinsky, Simon. "Nabokov and Chekhov: the lesser Russian tradition." Pp. 7–16 in Appel (3).

The Letters of John Keats, 1814–1821 2. ed., Hyder Edward Rollins Cambridge: Cambridge University Press, 1958.

Kenney, John Peter, ed. *The School of Moses: Studies in Philo and Hellenistic Religion*. Atlanta, Ga.: Scholars Press, 1995.

Kessler, Edward. *Coleridge's Metaphors of Being*. Princeton: Princeton University Press, 1979.

The Kingis Quair of James Stewart. ed. Matthew P. McDiarmid. London: Heinemann, 1973.

Kirk, G.S. "The Structure of the Homeric Hexameter." *Yale Classical Studies*, 20 (1966): 76–104.

Knowles, Owen and J.H. Stape. *Conrad: Intertexts & Appropriations: Essays in Memory of Yves Hervouet*. Amsterdam: Rodopi, 1997.

Knox, Israel. *The Aesthetic Theories of Kant, Hegel, and Schopenhauer*. New York: The Humanities Press, 1958.

Kohn, David, ed. *The Darwinian Heritage*. Princeton: Princeton University Press, 1985.

Kottler, Malcolm Jay (1). "Charles Darwin and Alfred Russel Wallace: Two Decades of Debate over Natural Selection." Pp. 367–432 in Kohn.

—— (2). "Alfred Russel Wallace." *Isis*, 65 (1974): 144–192.

Krabbe, Judith K. *The Metamorphoses of Apuleius*. New York: Lang, 1989.

Kratzmann, Gregory. *Anglo-Scottish Literary Relations 1430–1550*. Cambridge: Cambridge University Press, 1980.

Kuhn, Thomas. *The Structure of Scientific Revolutions*. Chicago: University of Chicago Press, 1996.

Kuzmanovich, Zoran. "'Splendid Insincerity' as 'Utmost Truthfulness': Nabokov and the Claims of the Real." Pp. 26–46 in Grayson (1).

LaFargue, J. Michael. *Language and Gnosis: the opening scenes of the Acts of Thomas*. Philadelphia: Fortress Press, 1985.

Lampert, Laurence. *Nietzsche's Teaching: An Interpretation of Thus Spake Zarathrustra*. New Haven: Yale University Press, 1986.

Langland, William. *Piers Plowman*. ed. A.V.C. Schmidt. Oxford: Oxford University Press, 1992.

Lanham, Urless Norton. *The Insects*. New York: Columbia University Press, 1964.

Larson, Edward J. *Evolution's Workshop: God and Science on the Galápagos Islands*. Harmondsworth: Allen Lane, 2001.

Lawrence, D.H. (1). *Apocalypse and the Writings on Revelation*. ed. Mara Kalnins. Cambridge: Cambridge University Press, 1980.

—— (2). *Reflections on the Death of a Porcupine and other essays*. ed. Michael Herbert. Cambridge: Cambridge University Press, 1988.

Layton, Robert, ed. *A Guide to the Symphony*. Oxford: Oxford University Press, 1995.

Leibniz, G.W. "Principles of Nature and of Grace, Based on Reason." *Philosophical Papers and Letters* 2. ed. Leroy E. Loemker. Dordrecht: Reidel, 1976.

Lesser, Friedrich Christian. *Théologie des insectes, ou, Demonstration des perfections de Dieu*. La Haye: Swart, 1742.

Levin, Simon Asher, ed. *Encyclopedia of Biodiversity* 5. San Diego: Academic Press, 2000.

Levine, Suzanne Jill. "Afterwords on Afterthoughts." *Review of Contemporary Fiction*, 10 no. 3 (Fall 1990): 181–82.

Lewis, John, ed. *Beyond Chance and Necessity: A Critical Inquiry into Professor Jacques Monod's 'Chance and Necessity.'* London: Garnstone Press, 1974.

Lewontin, Richard. *It Ain't Necessarily So: the dream of the human genome and other illusions*. London: Granta, 2000.

Livio, Mario. *The Golden Ratio: the Story of Phi, the Extra ordinary Number of Nature, Art and Beauty*. London: Review, 2003.

Londey, David and Carmen Johanson, eds. *Peri Hermencias: The Logic of Apuleius*. Leiden: Brill, 1987.

Lorentz, H.A., A. Einstein, H.M. and H. Weyl, eds. *The Principle of Relativity*. New York: Dover, 1952.

Loux, Michael J. and Dean W. Zimmerman *The Oxford Handbook of Metaphysics*. Oxford: Oxford University Press, 2003.

Luke, F.D. "Kafka's 'Die Verwandlung.'" *MLR* 46 (1951): 232–45.

Lumsden, Charles J. and Edward O. Wilson (1). *Genes, Mind and Culture*. Cambridge, Mass.: Harvard University Press, 1981.

—— (2). *Promethean Fire: Reflections on the Origin of Mind*. Cambridge, Mass.: Harvard University Press, 1983.

MacDiarmid, Hugh. *In Memoriam James Joyce: A Vision of World Language*. Glasgow: McLellan, 1955.

Machlup, F. and U. Mansfield, eds. *The Study of Information: Interdisciplinary Messages*. New York: Wiley, 1983.

Mackie, J.L. "Genes and Egoism." *Philosophy*, 56 (1981): 553–55.

MacQueen, John. *Numerology: theory and outline history of a literary mode*. Edinburgh: Edinburgh University Press, 1985.

Maeterlinck, Maurice (1). *Before the Great Silence*. London: Allen & Unwin, 1935.

—— (2). *The Life of the White Ant*. London: Allen & Unwin, 1927.

Magee, Bryan (1). *Misunderstanding Schopenhauer*. London: Institute of German Studies, 1990.

—— (2).*The Philosophy of Schopenhauer*. Oxford: Clarendon, 1983.

Magueijo, João. *Faster than the Speed of Light: the story of a scientific speculation*. London: Heinemann, 2003.

Maldacena, Juan. "Into the fifth dimension." *Nature*, 423 (12 June 2003): 695–96.

Mallet, James L. B. (1)."Concepts of Species." Levin: 427–40.

—— (2). "The genetics of biological diversity: from varieties to species." Pp. 13–53 in Gaston.

—— (3). "Mimicry meets the mitochondrion." *Current Biology*, 6, no.8: 937–940.

—— (4). "Poulton, Wallace and Jordan: how discoveries in *Papilio* butterflies initiated a new species concept 100 years ago" James Mallet, *Systematics and Biodiversity* (2004), prepublication: 1–14.

Mansfield, Jaap. "Heraclitus on the Psychology and Physiology of Sleep and Rivers." *Mnemosyne*, ser. 4, vol. 20 (1967): 1–29.

Marais, Eugene N. *The Soul of the White Ant*. London: Methuen, 1939.

Marcel, Gabriel. *Coleridge et Schelling*. Paris: Aubier-Montaigne, 1971.

Marchant, James, ed. *Alfred Russel Wallace: Letters & Reminiscences*. London: Cassell, 1916.

Margulis, Lynn. *The Symbiotic Planet: a New Look at Evolution*. New York: Basic Books, 1998.

Margulis, Lynn and Dorion Sagan. *Acquiring Genomes: a theory of the origins of species*. New York: Basic Books, 2002.

Márquez, Gabriel García. *Leaf Storm and other stories*. London: Cape, 1972.

Martner, Knud. *Selected Letters of Gustav Mahler*. London: Faber, 1979.

Mason, Bobbie Ann. *Nabokov's Garden: A Guide to Ada*. Ann Arbor: Ardis, 1974.

Matthiessen, F.O. *American Renaissance*. New York: Oxford University Press, 1941.

Maudlin, Tim. "Distilling Metaphysics from Quantum Theory." Pp. 461–87 in Loux.

Mayr, Ernst. *Evolution and the Diversity of Life: Selected Essays*. Cambridge, Mass.: Harvard University Press, 1997.

McFarland, Thomas. *Coleridge and the Pantheist Tradition*. Oxford: Clarendon, 1969.

McHale, Brian. *Postmodernist Fiction*. London: Routledge, 1989.

McIntosh, James. *Thoreau as Romantic Naturalist: His Shifting Stance toward Nature*. Ithaca: Cornell University Press, 1974.

McKinney, H. Lewis. *Wallace and Natural Selection*. New Haven: Yale University Press, 1972.

Mellers, Wilfrid. *The Sonata Principle*. London: Barrie & Jenkins,1973.

Melville, Herman (1). *Billy Budd and other Prose Pieces*. ed. Raymond Weaver. New York: Russell & Russell, 1963.

—— (2). *The Letters*. ed. Merrell R. Davis and William H. Gilman. New Haven 1960.

—— (3). *Mardi.* ed. Harrison Hayford, Hershel Parker and G. Thomas Tanselle. Evanston: Northwestern University Press, 1975.

—— (4). *Moby-Dick; or, The Whale.* ed. Harrison Hayford, Hershel Parker and G. Thomas Tanselle. Evanston: Northwestern University Press, 1988.

—— (5). *Pierre.* ed. Harrison Hayford, Hershel Parker and G. Thomas Tanselle. Evanston: Northwestern University Press, 1971.

Merkelbach, Reinhold. *Roman und Mysterium in der Antike.* Munich: Beck, 1962.

Michel, Paul Henri. *De Pythagore á Euclide: contribution à l'histoire des mathématiques préeucliennes* (Paris: Collection des études anciennes, 1950)

Midgley, Mary. "Gene-juggling." *Philosophy,* 54 (1979): 439–58.

Miller, George A. "The Magical Number Seven, Plus or Minus Two: Some Limits on Our Capacity for Processing Information." *Psychological Review,* 63 (1956): 81–97.

Moehring, Horst R. "Arithmology as an Exegetical Tool in the Writing of Philo of Alexandria." Pp. 141–76 in Kenney.

Mommsen, Theodor. *The History of Rome.* London: Routledge/ Thoemmes Press, 1996.

Moore, Maxine. *That Lonely Game.* Columbia: University of Missouri Press, 1957.

Moore, Walter, *Schrödinger: Life and Thought.* Cambridge: Cambridge University Press, 1989.

Moravia, Alberto. *Man as an End: a defense of humanism.* London: Secker & Warburg, 1965.

Morgan, Bayard Q. "On the Use of Number in the *Nibelungenlied.*" *Journal of English and Germanic Philology,* 36 (1937): 10–20.

Morris, Ian and Barry B. Powell. *A New Companion to Homer.* Leiden: Brill, 1997.

Moszkowski, Alexander. *Conversations with Einstein.* trans. Henry L. Brose. London: Sidgwick & Jackson, 1972.

Moyal, George J.D. *René Descartes: Critical Assessments.* London: Routledge & Kegan Paul, 1991.

Muir, Edwin (1). *The Complete Poems.* ed. Peter Butter. Aberdeen: ASLS, 1991.

—— (2). *Essays on Literature and Society.* London: Hogarth, 1949.

—— (3). *Latitudes.* New York: Huebsch, 1924.

—— (4). *The Story and the Fable.* London: Harrap, 1940.

Muir, John (1). *John of the Mountains: the Unpublished Journals of John Muir* ed. Linnie Marsh Wolfe. Madison: Wisconsin University Press, 1979.

—— (2) *The Wilderness Journey.* Edinburgh: Canongate, 1996.

Muirhead, John Henry. *Coleridge as Philosopher.* London: Allen & Unwin, 1930.

Murray, Henry A. "In Nomine Diaboli." Pp. 62–74 in Chase.

Murray, J.D. "On pattern formation mechanisms for lepidopteran wing patterns and mammalian coat markings." Pp. 47–70 in Brenner.

Murry, J. Middleton. *William Blake.* London: Cape, 1933.

The Nag Hammadi Library in English. Leiden: E.J. Brill, 1977.

Nabokov, Vladimir (1). *Ada or Ardor: a Family Chronicle.* London: Weidenfeld & Nicolson, 1969.

—— (2). *The Annotated Lolita.* ed. Alfred Appel, Jr. Weidenfeld & Nicolson, 1971)

—— (3). *Bend Sinister*. Harmondsworth: Penguin, 2001.

—— (4). *The Gift*. Harmondsworth: Penguin, 1981.

—— (5). *Invitation to a Beheading*. London: Weidenfeld & Nicolson, 1960.

—— (6). *King, Queen, Knave*. Weidenfeld & Nicolson, 1968.

—— (7). *Lectures on Literature*. London: Weidenfeld & Nicolson, 1980.

—— (8). *Mary*. London: Weidenfeld & Nicolson, 1971.

—— (9). *Nabokov's Dozen*. Harmondsworth: Penguin, 1984.

—— (10). *Nikolay Gogol*. London: Weidenfeld & Nicolson, 1973.

—— (11). *Poems and Problems*. London: Weidenfeld & Nicolson, 1972.

—— (12). *The Real Life of Sebastian Knight*. Harmondsworth: Penguin, 1982.

—— (13). *Speak Memory: An Autobiography Revisited*. London: Everyman, 1999.

—— (14). *The Stories of Vladimir Nabokov*. New York: Vintage, 1997.

—— (15). *Strong Opinions*. London: Weidenfeld & Nicolson, 1974.

Nash, Suzanne. *Les Contemplations of Victor Hugo*. Princeton: Princeton University Press, 1976.

Nehring, A. "Homer's Descriptions of Syncopes" *Classical Philology* 42 (1947): 106–21.

Nerlich, Graham."Space-Time Substantivalism." Pp. 281–314 in Loux.

Newell, A. "Intellectual issues in the History of Artificial Intelligence." Machlup: 196–227.

Nicholson, A.J. *Evolution after Darwin*. ed. S.Tax. Chicago: Chicago University Press, 1960.

Nicomachus. *Introduction to Arithmetic*. trans. Martin Luther D'Ooge. New York: University of Michigan Studies, 1926.

Nietzsche, F. *Complete Works*. ed. Oscar Levy. Edinburgh: Foulis, 1909–13.

Nijhout, H. Frederick (1). *The Development and Evolution of Butterfly Wing Patterns*. Washington: Smithsonian Institution Press, 1991.

—— (2). "The Color Patterns of Butterflies and Moths," *Scientific American* (November 1981): 104–15.

Nuttall, A.D. *The Alternative Trinity: Gnostic Heresy in Marlowe, Milton, and Blake*. Oxford: Clarendon, 1998.

O'Brien, Flann. *At Swim-Two-Birds*. Harmondsworth: Penguin, 1967.

Oldroyd, D.R. *Darwinian Impacts: an introduction to the Darwinian Revolution*. Milton Keynes: Open University Press, 1980.

O'Meara, Dominic J. *Pythagoras Revived: Mathematics and Philosophy in Late Antiquity*. Oxford: Clarendon, 1989.

Onians, Richard Broxton. *The Origin of European Thought: about the body, the mind, the soul, the world, time, and fate*. Cambridge: Cambridge University Press, 1951.

Ouspensky, Peter Demianovich. *A New Model of the Universe: principles of the psychological method in its application to problems of science, religion, and art*. London: Routledge & Kegan Paul, 1938.

Ovid. *Metamorphoses*. trans. Mary M. Innes. Harmondsworth: Penguin, 1955.

Paley, Morton D. *The Continuing City: William Blake's Jerusalem*. (Oxford: Oxford University Press, 1983).

Palmquist, Stephen R. *Kant's System of Perspectives: an architectonic interpretation of the critical philosophy*. Lanham, Md.: University Press of America, 1993.

Panofsky, Erwin. *Studies in Iconology: Humanistic Themes in the Art of the Renaissance*. New York: Icon, Harper, 1972.

Pascal, Roy. *Kafka's Narrators: a study of his stories and sketches*. Cambridge: Cambridge University Press, 1982.

Patrides, C.A. "The Numerological Approach to Cosmic Order during the English Renaissance." *Isis*, 49, pt.4, no. 158 (December 1958): 391–97.

Paulhan, Jean (1). *La preuve par l'étymologie*. Cognac: Le temps qu'il fait, 1988.

—— (2)."Jacob Cow the Pirate or Whether Words are Signs." Pp. 113–24 in *Essays in Language and Literature*, edited by J.L. Hevesi. London: Allan Wingate, 1947.

Paz, Octavio. "The Power of Ancient Mexican Art." *New York Review of Books* 37 no. 19 (December 6, 1990): 18–21.

Penrose, Roger. *The Road to Reality: a complete guide to the laws of the universe*. London: Cape 2004.

Perry, Ben Edwin (1). *The Ancient Romances*. Berkeley and Los Angeles: University of California, 1967.

—— (2). "An Interpretation of Apuleius' *Metamorphoses*." *TAPhA* 57 (1926): 238–60.

Pinker, Steven. *How the Mind Works*. London: Allen Lane, 1997.

Pinto, Vivian de Sola (1). *The Divine Vision: studies in the poetry and art of William Blake*. London: Gollancz, 1957.

Pinto, Vivian de Sola (2) and Warren Roberts eds. *The Complete Poems of D.H. Lawrence*. London: Heinemann, 1967.

Plato (1). *The Cratylus, Phaedo, Parmenides, and Timaeus of Plato*. trans. Thomas Taylor. London: B. & J. White, 1793.

—— (2). *Phaedrus*. ed. H,N. Fowler. London: Heinemann, 1926.

Popkin, Richard H., ed. *The Columbia History of Western Philosophy*. New York: Columbia University Press, 1999.

Porte, Joel. *Emerson in his Journals*. Cambridge, Mass.: Belknap Press, 1982.

Porter, Carolyn. "Call Me Ishmael, or How to Make double-Talk Speak." Pp. 73–108 in Brodhead.

Powell, Barry B. *Homer and the origins of the Greek alphabet*. Cambridge: Cambridge University Press, 1991.

—— (2). *Writing and the Origins of Greek Literature*. Cambridge: Cambridge University Press, 2002.

Punnett, Reginald Crundall. *Mimicry in Butterflies*. Cambridge: Cambridge University Press, 1915.

Purser, Louis C. *The Story of Cupid and Psyche as related by Apuleius*. New York: Caratzas, 1983.

Pyenson, Lewis. *The Young Einstein: the advent of relativity*. Bristol: Hilger, 1985.

Quinby, Lee. *Freedom, Foucault, and the Subject of America*. Boston: Northeastern University Press, 1991.

Qvarnström, Gunnar. *The Enchanted palace: Some Structural Aspects of Paradise Lost*. Stockholm: Almqvist & Wiksell, 1967.

Raby, Peter. *Alfred Russel Wallace: A Life*. London: Chatto & Windus, 2001.

Raine, Kathleen (1). *Blake and Tradition*. London: Routledge & Kegan Paul, 1969.

—— (2). *Golgonooza City of Imagination: Last Studies in William Blake*. Ipswich: Golgonooza Press, 1991.

Rasnitsyn, Alexander P. and Donald L.J. Quicke, eds. *History of Insects*. Dordrecht: Kluwer, 2002.

Rea, Michael. "Four-Dimensionalism." Pp. 246–80 in Loux.

Read, Herbert. *The True Voice of Feeling: Studies in English Romantic Poetry*. London: Faber, 1947.

Resh, Vincent H. and Ring T. Carde, eds. *Encyclopedia of Insects*. Orlando: Academic Press, 2003.

Rees, Martin. *Just Six Numbers: the Deep forces that Shape the Universe*. London: Weidenfeld & Nicolson, 1999.

Reiss, Edmund. "Number Symbolism and Medieval Literature." *Medievalia et Humanista*, n.s. 1 (1970): 161–74.

Reitzenstein, Richard the Elder. *Das Märchen von Amor und Psyche bei Apuleius*. Leipzig: University of Freiburg,1911.

Richards, O.W., and R.G. Davies. *Imms' General Textbook of Entomology*. 10th ed. London: Chapman & Hall, 1994.

Ríos, Julián. *Larva: Midsummer Night's Babel*. trans. Richard Alan Francis. London: Quartet, 1991.

Robayna, Andrés Sánchez, ed. *Palabras para Larva*. (Barcelona: Edicions del Mall, 1985).

Robbins, Frank Eggleston. "The Tradition of Greek Arithmology." *Classical Philology*, 16, no. 2 (April 1921): 97–123.

Robinson, D.W., Jr., and Bernard F. Huppé. *Piers Plowman and Scriptural Tradition*. Princeton: Princeton University Press, 1951.

Robson, C.A. "Quatrains and Passages of 8 Lines in Béroul: Some Stylistic and Linguistic Aspects." Pp. 171–201 in Barnett.

Rohde, Erwin. *Psyche: The Cult of Souls and Belief in Immortality among the Greeks*. London: Kegan Paul, Trench, Trubner, 1925.

Roques, Mario (1). "Notes Pour L'Edition de la *Vie de saint Grégoire* en Ancien Français," *Romania*, 78 (1956): 1–25.

—— (2). "Sur deux particularités métriques de la *Vie de saint Grégoire* en ancien français." *Romania*, 48 (1922): 41–61.

Rose, E.J. "'Mental Forms Creating': 'Fourfold Vision' and the Poet as Prophet in Blake's Designs and Verse" *Journal of Aesthetics and Art Criticism*. 12 (1964): 173–83.

Rosen, Charles. *The Classical Style: Haydn, Mozart, Beethoven*. London: Faber, 1976.

Rosen, Stanley *The Ancients and the Moderns: rethinking modernity*. New Haven: Yale University Press, 1989.

Rosenfield, Leonora Cohen, *From Beast-Machine to Man-Machine: Animal soul in French Letters from Descartes to La Mettrie*. New York: Octopus, 1968.

Ross, Kelley, L. *Arthur Schopenhauer*. <http://www.friesian.com/arthur.htm> (1998).

Röstvig, Maren-Sofia. "Structure as prophecy: the influence of biblical exegesis upon theories of literary structure." Pp. 32–72 in Fowler (2).

Ruse, Michael. *Sociobiology: Sense or Nonsense.* Dordrecht: Reidel, 1979.

Rushdie, Salman. *Imaginary Homelands.* London: Granta, 1991.

Russell, Bertrand. *A Critical Exposition of the Philosophy of Leibniz.* London: Allen & Unwin, 1958.

Ruxton, G.D., M. Speed and T.N. Sherratt. "Evasive mimicry: when (if ever) could mimicry based on difficulty of capture evolve?" *Proceedings of the Royal Society of London* B 271 (2004): 2135–42.

Salinas, Pedro. *Reality and the Poet in Spanish Poetry.* Baltimore: Johns Hopkins, 1940.

Schelling, F. W. (1). *Ideas for a Philosophy of Nature: as Introduction to the Study of this Science.* 2nd ed. 1803. Ed. Errol E. Harris and Peter Heath. Introduction by Robert Stern. Cambridge: Cambridge University Press, 1988.

—— (2). *On the History of Modern Philosophy.* ed., Andrew Bowie. Cambridge: Cambridge University Press, 1994.

—— (3).*The Philosophy of Art.* ed. and trans. Douglas W. Scott. Foreword by David Simpson. Minneapolis: University of Minnesota Press, 1989.

—— (4). *System of Transcendental Idealism.* Charlottesville: University Press of Virginia, 1978.

Schlam, Carl. *Cupid and Psyche: Apuleius and the Monuments.* Pennsylvania: American Philological Association, 1976.

—— (2). *The Metamorphoses of Apuleius: on making an ass of oneself.* London: Duckworth, 1992.

Schlipp, Paul Arthur, ed. *Albert Einstein: Philosopher-Scientist.* Lasalle, Ill.: Open Court, 1949.

Scholem, Gershom G. *Major Trends in Jewish Mysticism.* New York: Schoken Books, 1961.

Schopenhauer, Arthur (1). *Manuscript Remains in 4 Volumes.* ed. Arthur Hübscher. trans. E.F.J. Payne. Oxford: Berg, 1988–90.

—— (2). *On the Will in Nature: A Discussion of the Corroborations from the Empirical Sciences That the Author's Philosophy has received since its first appearance.* trans. E.F.J. Payne. edited with a preface by David E. Cartwright. Oxford: Berg, 1992.

—— (3). *Parerga and Paralipomena: short philosophical essays.* trans. E.F.J. Payne. Oxford: Clarendon, 2000.

—— (4). *Schopenhauer's early Fourfold Root.* trans. and commentary by F.C. White. Aldershot: Avebury, 1997.

—— (5). *The World as Will and Representation.* trans. E.F.J. Payne. New York: Dover, 1958.

Scriber, J. Mark, Yoshitaka Tsubaki, Robert C. Lederhouse, eds. *Swallowtail Butterflies: their ecology and evolutionary biology.* Gainesville: Scientific Publishers, 1995.

Searle, John R. *The Mystery of Consciousness.* London: Granta, 1997.

Sherry, Norman. *Conrad's Eastern World.* Cambridge: Cambridge University Press, 1966.

Shumaker, Wayne. *John Dee on Astronomy: Propaedeumata Aphoristica (1558 & 1568)*. Berkeley and Los Angeles: University of California Press, 1978.

Sider, Theodore (1). *Four-Dimensionalism: an ontology of persistence and time*. Oxford: Oxford University Press, 2001.

—— (2). "Four Dimensionalism." *Philosophical Review*, 106 no. 2 (April 1997): 197–231.

Silkin, Jon. *The Life of Metrical and Free Verse in Twentieth-Century Poetry*. New York: St. Martin's Press, 1997.

Simpson, David, ed. *The Origins of Modern Critical Thought: German aesthetic and literary criticism from Lessing to Hegel*. Cambridge: Cambridge University Press, 1988.

Simpson, Robert. *Beethoven Symphonies*. London: BBC, 1970.

Singhal, D.P. *India and World Civilization*. London: Sidgwick & Jackson, 1972.

Singleton, Charles S., ed. *Art, Science and History in the Renaissance*. Baltimore: Johns Hopkins University Press, 1967.

Smith, Charles H., ed. *Alfred Russel Wallace: An Anthology of his Shorter Writings*. Oxford: Oxford University Press, 1991.

Smith, John Maynard (1). *The Theory of Evolution*. Harmondsworth: Pelican, 1975.

—— (2). *Did Darwin Get it Right? Essays on Games, Sex and Evolution*. Harmondsworth: Penguin, 1993.

—— (3). "Genes, Memes, & Minds." *New York Review of Books* (30 November, 1995): 46–48.

Smith, Roger. "Alfred Russel Wallace: Philosophy of Nature and Man." *British Journal for the History of Science*, 6: pt. 2, no. 22 (December 1972): 177–99.

Smolin, Lee. *Three Roads to Quantum Gravity: A New Understanding of Space, Time and the Universe*. London: Weidenfeld & Nicolson, 2000.

Snell, Bruno. *The Discovery of the Mind in Greek philosophy and literature*. New York: Dover, 1982.

Sokel, Walter H. "Kafka's 'Metamorphosis': Rebellion and Punishment." *Monatschefte* 48 no. 4 (1956).

Sorensen, Peter J. *William Blake's Recreation of Gnostic Myth: resolving the apparent incongruities*. Lewiston: Mellen Press, 1995.

Southern, R.W. *Medieval Humanism and other studies*. Oxford: Oxford University Press, 1970.

Spengler, Oswald. *The Decline of the West*. London: Allen & Unwin, 1926–28.

Speyer, Edward. *Six Roads from Newton: Great Discoveries in Physics*. New York: Wiley, 1994.

Spooner, David (1). *The Metaphysics of Insect Life and other essays*. San Francisco: International Scholars, 1995.

—— (2). *The Poem and the Insect: aspects of twentieth century Hispanic culture*. Lanham, Md.: University Press of America, 1999, 2002).

Stanford, W.B. *The Sound of Greek: studies in the Greek theory and practice of euphony*. Berkeley and Los Angeles: University of California, 1967.

Steiner, George (1). *After Babel: aspects of language and translation*. Oxford: Oxford University Press, 1975.

—— (2). *Extraterritorial: papers on literature and the language revolution*. Harmondsworth: Penguin, 1975.

—— (3). *Grammars of Creation*. London: Faber, 2001.

Sterelny, Kim. *Dawkins vs. Gould: survival of the fittest*. Cambridge: Icon Books, 2001.

Stewart, Ian (1). *Life's Other Secret: the new mathematics of the living world*. London: Allen Lane, 1998.

—— (2). *Nature's Numbers: Discovering Order and Pattern in the Universe*. Weidenfeld & Nicolson, 1995.

Stewart, Ian (3) and Martin Golubitsky. *Fearful Symmetry: is God a geometer?* London: Penguin, 1993.

Susskind, Leonard. "A universe like no other." *New Scientist*, 180: 2419 (1 November, 2003): 34–41.

Swedenborg, Emanuel *The Earths in the Universe*. London: Swedenborg Society, 1894. In the Wallace Collection, Edinburgh University Library. Read and marked by A.R. Wallace, 1903.

—— (2). *God, Creation, Man*. London: Warne, 1905. In the Wallace Collection, Edinburgh University Library. Read and marked by A.R. Wallace, n.d.

Tanner, Michael. *Schopenhauer: Metaphysics and Art*. London: Phoenix, 1998.

Tax, S., ed. *Evolution after Darwin* 1. Chicago: Chicago University Press, 1960.

Taylor, A.E. "Forms and Numbers: A Study in Platonic Metaphysics." *Mind*, 35 (1926): 419–40, and 36 (1927): 12–33.

Taylor, John, "Gauge theories in particle physics." In Davies (3).

Thomas, J.A. et al. "Comparative Losses of British Butterflies, Birds, and Plants and the Global Extinction Crisis." *Science* 303 (19 March 2004): 1879–81.

Thompson, Lawrance. *Melville's Quarrel with God*. Princeton: Princeton University Press, 1952.

Thoreau, H.D. (1). *Excursions*. Boston: Ticknor & Fields, 1863.

—— (2). *Journal*. ed. Elizabeth Hall Witherell et al. Princeton: Princeton University Press, 1981–. 7 volumes.

—— (3). *The Journal of Henry David Thoreau*. ed. Bradford Torrey and Francis Allen. New York: Dover, 1962.

—— (4). *Walden*. ed. J. Lyndon Shanley. Princeton: Princeton University Press, 1989.

—— (5). *A Week on the Concord and Merrimack Rivers*. ed. Carl F. Hovde, William L. Howarth and Elizabeth Hall Witherell. Princeton: Princeton University Press, 1980.

Tindall, William York. *A Readers Guide to Finnegans Wake*. London: Thames & Hudson, 1969.

Truman, James W. and Lynn M. Riddiford. "The Origins of Insect Metamorphosis." *Nature*, 41 (30 September 1999): 447–52.

Tsuchiya, Kiyoshi. *The mirror metaphor and Coleridge's mysticism: poetics, metaphysics, and the formation of the pentad*. Lewiston: Mellen Press, 2000.

Turnell, Martin. *The Classical Moment: studies of Corneille, Molière and Racine*. London: Hamish Hamilton, 1947.

Turner, John R.G. "Why male butterflies are non-mimetic: natural selection, sexual selection, group selection, modification and sieving." *Biological Journal of the Linnean Society* (December 1978), 10: 385–432.

Vorzimmer, Peter J. *Charles Darwin: The Years of Controversy: the Origin of Species and its critics 1859–82*. Philadelphia: Temple University Press, 1972.

Wace, Alan J.B. and Frank H. Stubbings. *A Companion to Homer*. London: Macmillan, 1962.

Wade, Joseph Sanford. "The Friendship of Two Old-Time Naturalists." *The Scientific Monthly*, 23 (August 1926): 152–60.

Walker, James R. *Lakota Belief and Ritual*. Lincoln: University of Nebraska Press, 1980.

Wallace, Alfred Russel (1). *Contributions to the Theory of Natural Selection*. London: Macmillan, 1871.

——— (2). *Darwinism: an exposition of the Theory of Natural Selection with some of its applications*. London: Macmillan, 1889.

——— (3). *Island Life, or the phenomena and causes of Insular Faunas and Floras: including a revision and attempted solution of the problem of geological climates*. London: Macmillan, 1892.

——— (4). *The Malay Archipelago: the land of the organg-utan and the bird of paradise. A narrative of travel, with studies of man and nature*. 5th ed. London: Macmillan, 1874).

——— (5). *My Life: a record of events and opinions*. London: Chapman & Hall, 1905.

——— (6). *A Narrative of Travels on the Amazon and Rio Negro: with an account of the native tribes, and observations on the climate, geology, and natural history of the Amazon Valley*. London: Ward, Lock & Co., 1890.

——— (7). *Natural Selection and Tropical Nature: Essays on Descriptive and Theoretical Biology*. London: Macmillan, 1891.

——— (8). *The Wonderful Century: its successes and its failures*. London: Swan Sonnenschein, 1898.

——— (9). *The World of Life: A Manifestation of Creative Power, Directive Mind and Ultimate Purpose*. London: Chapman & Hall, 1910.

——— (10). "Description of a New Species of Ornithoptera. Ornithoptera brookiana." *Proceedings of the Entomological Society of London* (1854–55): 104–05.

——— (11). "Geological Climates and the Origin of Species." *Quarterly Review*, 126 (1869): 391–94. Also in Wallace (7).

——— (12). "Limits of Natural Selection as applied to man." In Wallace (1).

——— (13), "Mimicry and other Protective Resemblances Among Animals," *Westminster Review*, 32 (n.s.), no. 1: 1–43.

——— (14)."Natural Selection." *Athenaeum*, no. 2040 (1 December 1866): 716–17.

——— (15). "On the Entomology of the Aru Islands." *Zoologist* 16 (January 1858), nos. 185–86: 5889–94.

——— (16). "On the Habits of the Butterflies of the Amazon Valley." *Transactions of the Entomological Society of London*, 2 (n.s.): pt. 8 (April 1854): 253–64.

——— (17). "On the Law Which Has Regulated the Introduction of New Species." *Annals and Magazine of Natural History* 16 (2nd ser.) (September 1855): 184–96.

—— (18). "On the Tendency of Varieties to Depart Indefinitely from the Original Type" *Journal of the Proceedings of the Linnean Society: Zoology* 3 (9) (20 August 1858): 45–62.

—— (19). "The Phenomena of Variation and Geographical Distribution as Illustrated by the Papilionidae of the Malayan Region." *Transactions of the Linnean Society of London*, 25: pt 1 (1865): 1–71.

—— (20). "Protective Mimicry in Animals." in *Science for All* 2, ed. Robert Brown (London: Cassell, Petter, Galpin & Co., 1879): 284–96.

—— (21). "Regarding Mimicry in Insects." *Journal of the Proceedings of the Entomological Society of London*, (1864): 14–15.

Walsh, P.G. *The Roman Novel*. London: Bristol Classical, 1995.

Watson, James D. *Letter to the Author*, 17 April 1980.

Wedberg, Anders. *Plato's Philosophy of Mathematics*. Stockholm: Almqvist & Wiksell, 1955.

Weinberg, Steven. *Dreams of a Final Theory*. London: Hutchinson Radius, 1993.

Wenzl, Aloys. "Einstein's Theory of Relativity, viewed from the standpoint of its critical realism, and its significance for philosophy." Pp. 581–606 in Schlipp.

West, Geoffrey B., James H. Brown, Brian J. Enquist. "A General Model for the Origin of Allometric Scaling Laws in Biology." *Science* 276:5309 (April 4, 1997): 122–26.

West, Nathanael. *Complete Works*. London: Picador, 1983.

Westcott, W. Wynn. *Numbers: their occult power and mystic virtues*. London: Theosophical Society, 1911.

Whitfield, Philip. *Evolution: the greatest story ever told*. London: Marshall, 1998.

Willey, Basil. *Darwin and Butler: two versions of evolution*. London: Chatto & Windus, 1960.

Williams-Ellis, Amabel. *Darwin's Moon: a biography of Alfred Russel Wallace*. London: Blackie, 1966.

Williamson, Donald I. *Larvae and Evolution: toward a new zoology*. London: Chapman & Hall, 1992.

Wind, Edgar. *Pagan Mysteries in the Renaissance*. Oxford: Oxford University Press, 1989.

Wilson, Edward O. (1). *On Human Nature*. Cambridge, Mass.: Harvard University Press, 1978.

—— (2). *Sociobiology: the new synthesis*. Cambridge, Mass.: Belknap, 1975.

Wilson, Mona. *The Life of William Blake*. London: Hart-Davis, 1948.

Wink, Walter. *Cracking the Gnostic Code: the Powers of Gnosticism*. Atlanta: Scholars Press, 1993.

Wood, Rupert. "Language as Will and Representation: Schopenhauer, Austin, and Musicality." *Comparative Literature* 48 no. 4 (Fall 1996): 302–25.

Woodard, Roger D. *Greek Writing from Knossos to Homer: A Linguistic Interpretation of the Origin of the Greek Alphabet and the Continuity of Ancient Greek Literacy*. Oxford: Oxford University Press, 1997.

Woodcock, George. *W.H. Bates: Naturalist of the Amazons*. London: Faber, 1969.

Yeats, W.B. (1). *A Critical Edition of Yeats's 'A Vision.'* ed. George Mills Harper and Walter Kelly Hood. London: Macmillan, 1978.

—— (2). *A Vision and Related Writings*. ed. A. Norman Jeffares. London: Arena, 1990.

—— (3). *A Vision*. London: T. Werner Laurie, 1925.

Young, Mark. *The Natural History of Moths*. London: T. & A.D. Poyser, 1997.

Yourgrau, Palle. *The disappearance of time: Kurt Gödel and the Idealistic Tradition in Philosophy*. Cambridge: Cambridge University Press, 1991.

Zimmer, Dieter. "Mimicry in Nature and Art." Pp. 47–57 in Grayson (1).

Index